THE WILEY GUIDE TO PROJECT CONTROL

THE WILEY GUIDES TO THE MANAGEMENT OF PROJECTS

Edited by

Peter W. G. Morris and Jeffrey K. Pinto

The Wiley Guide to Project, Program & Portfolio Management
978-0-470-22685-8

The Wiley Guide to Project Control
978-0-470-22684-1

The Wiley Guide to Project Organization & Project
Management Competencies
978-0-470-22683-4

The Wiley Guide to Project Technology, Supply Chain &
Procurement Management
978-0-470-22682-7

THE WILEY GUIDE TO PROJECT CONTROL

Edited by

Peter W. G Morris and Jeffrey K. Pinto

JOHN WILEY & SONS, INC.

This book is printed on acid-free paper. ♾

Copyright © 2007 by John Wiley & Sons, Inc. All rights reserved.

Published by John Wiley & Sons, Inc., Hoboken, New Jersey.
Published simultaneously in Canada.

Wiley Bicentennial Logo: Richard J. Pacifico

No part of this publication may be reproduced, stored in a retrieval system, or transmitted in any
form or by any means, electronic, mechanical, photocopying, recording, scanning, or otherwise,
except as permitted under Section 107 or 108 of the 1976 United States Copyright Act, without
either the prior written permission of the Publisher, or authorization through payment of the
appropriate per-copy fee to the Copyright Clearance Center, 222 Rosewood Drive, Danvers, MA
01923, (978) 750-8400, fax (978) 646-8600, or on the web at www.copyright.com. Requests to the
Publisher for permission should be addressed to the Permissions Department, John Wiley & Sons,
Inc., 111 River Street, Hoboken, NJ 07030, (201) 748-6011, fax (201) 748-6008, or online at
www.wiley.com / go / permissions.

Limit of Liability / Disclaimer of Warranty: While the publisher and the author have used their best
efforts in preparing this book, they make no representations or warranties with respect to the
accuracy or completeness of the contents of this book and specifically disclaim any implied warranties
of merchantability or fitness for a particular purpose. No warranty may be created or extended by
sales representatives or written sales materials. The advice and strategies contained herein may not be
suitable for your situation. You should consult with a professional where appropriate. Neither the
publisher nor the author shall be liable for any loss of profit or any other commercial damages,
including but not limited to special, incidental, consequential, or other damages.

For general information about our other products and services, please contact our Customer Care
Department within the United States at (800) 762-2974, outside the United States at (317) 572-3993
or fax (317) 572-4002.

Wiley also publishes its books in a variety of electronic formats. Some content that appears in print
may not be available in electronic books. For more information about Wiley products, visit our web
site at www.wiley.com.

Library of Congress Cataloging-in-Publication Data:

ISBN: 978-0-470-22684-1

Printed in the United States of America

10 9 8 7 6 5 4 3 2 1

CONTENTS

THE WILEY GUIDE TO PROJECT CONTROL:
PREFACE AND INTRODUCTION

Peter W. G. Morris and Jeffrey Pinto

In 1983, Dave Cleland and William King produced for Van Nostrand Reinhold (now John Wiley & Sons) the *Project Management Handbook*, a book that rapidly became a classic. Now over twenty years later, Wiley is bringing this landmark publication up to date with a new series *The Wiley Guides to the Management of Projects*, comprising four separate, but linked, books.

Why the new title—indeed, why the need to update the original work?

That is a big question, one that goes to the heart of much of the debate in project management today and which is central to the architecture and content of these books. First, why "the management of projects" instead of "project management"?

Project management has moved a long way since 1983. If we mark the founding of project management to be somewhere between about 1955 (when the first uses of modern project management terms and techniques began being applied in the management of the U.S. missile programs) and 1969/70 (when project management professional associations were established in the United States and Europe) (Morris, 1997), then Cleland and King's book reflected the thinking that had been developed in the field for about the first twenty years of this young discipline's life. Well, over another twenty years has since elapsed. During this time there has been an explosive growth in project management. The professional project management associations around the world now have thousands of members—the Project Management Institute (PMI) itself having well over 200,000—and membership continues to grow! Every year there are dozens of conferences; books, journals, and electronic publications abound; companies continue to recognize project management as a core business discipline and work to improve company performance through it; and, increasingly, there is more formal educational work carried out in university teaching and research programs, both at the undergraduate, and particularly graduate, levels.

Yet, in many ways, all this activity has led to some confusion over concepts and applications. For example, the basic American, European, and Japanese professional models of project management are different. The most influential, PMI, not least due to its size, is the most limiting, reflecting an essentially execution, or delivery, orientation, evident both in its *Guide to the Project Management Body of Knowledge, PMBOK Guide, 3rd Edition* (PMI, 2004) and its *Organizational Project Management Maturity Model, OPM3* (PMI, 2003). This approach tends to under-emphasize the front-end, definitional stages of the project, the stages that are so crucial to successful accomplishment (the European and Japanese models, as we shall see, give much greater prominence to these stages). An execution emphasis is obviously essential, but managing the definition of the project, in a way that best fits with the business, technical, and other organizational needs of the sponsors, is critical in determining how well the project will deliver business benefits and in establishing the overall strategy for the project.

It was this insight, developed through research conducted independently by the current authors shortly after the publication of the Cleland and King *Handbook* (Morris and Hough, 1987; Pinto and Slevin, 1988), that led to Morris coining the term "the management of projects" in 1994 to reflect the need to focus on managing the definition and delivery of *the project itself* to deliver a successful outcome.

These at any rate are the themes that we shall be exploring in this book (and to which we shall revert in a moment). Our aim, frankly, is to better center the discipline by defining more clearly what is involved in managing projects successfully and, in doing so, to expand the discipline's focus.

So second, why is this endeavor so big that it takes four books? Well, first, it was both the publisher's desire and our own to produce something substantial—something that could be used by both practitioners and scholars, hopefully for the next 10 to 20 years, like the Cleland and King book—as a reference for the best-thinking in the discipline. But why are there so many chapters that it needs four books? Quite simply, the size reflects the growth of knowledge within the field. The "management of projects" philosophy forces us (i.e., members of the discipline) to expand our frame of reference regarding what projects truly *are* beyond of the traditional *PMBOK/OPM3* model.

These, then, are not a set of short "how to" management books, but very intentionally, resource books. We see our readership not as casual business readers, but as people who are genuinely interested in the discipline, and who is seek further insight and information— the thinking managers of projects. Specifically, the books are intended for both the general practitioner and the student (typically working at the graduate level). For both, we seek to show where and how practice and innovative thinking is shaping the discipline. We are deliberately pushing the envelope, giving practical examples, and providing references to others' work. The books should, in short, be a real resource, allowing the reader to understand how the key "management of projects" practices are being applied in different contexts and pointing to where further information can be obtained.

To achieve this aim, we have assembled and worked, at times intensively, with a group of authors who collectively provide truly outstanding experience and insight. Some are, by any standard, among the leading researchers, writers, and speakers in the field, whether as academics or consultants. Others write directly from senior positions in industry, offering their practical experience. In every case, each has worked hard with us to furnish the relevance, the references, and the examples that the books, as a whole, aim to provide.

What one undoubtedly gets as a result is a range that is far greater than any individual alone can bring (one simply cannot be working in all these different areas so deeply as all these authors, combined, are). What one does not always get, though, are all the angles that any one mind might think is important. This is inevitable, if a little regrettable. But to a larger extent, we feel, it is beneficial for two reasons. One, this is not a discipline that is now done and finished—far from it. There are many examples where there is need and opportunity for further research and for alternative ways of looking at things. Rodney Turner and Anne Keegan, for example, in their chapter on managing innovation (*The Wiley Guide to Project Technology, Supply Chain & Procurement Management*, Chapter 8) ended up positioning the discussion very much in terms of learning and maturity. If we had gone to Harvard, to Wheelwright and Clark (1992) or Christensen (1999) for example, we would almost certainly have received something that focused more on the structural processes linking technology, innovation, and strategy. This divergence is healthy for the discipline, and is, in fact, inevitable in a subject that is so context-dependent as management. Second, it is also beneficial, because seeing a topic from a different viewpoint can be stimulating and lead the reader to fresh insights. Hence we have Steve Simister giving an outstandingly lucid and comprehensive treatment in *The Wiley Guide to Project Control*, Chapter 5 on risk management; but later we have Stephen Ward and Chris Chapman coming at the same subject (*The Wiley Guide to Project Control*, Chapter 6) from a different perspective and offering a penetrating treatment of it. There are many similar instances, particularly where the topic is complicated, or may vary in application, as in strategy, program management, finance, procurement, knowledge management, performance management, scheduling, competence, quality, and maturity.

In short, the breadth and diversity of this collection of work (and authors) is, we believe, one of the books' most fertile qualities. Together, they represent a set of approximately sixty authors from different discipline perspectives (e.g., construction, new product development, information technology, defense / aerospace) whose common bond is their commitment to improving the management of projects, and who provide a range of insights from around the globe. Thus, the North American reader can gain insight into processes that, while common in Europe, have yet to make significant inroads in other locations, and vice versa. IT project managers can likewise gather information from the wealth of knowledge built up through decades of practice in the construction industry, and vice versa. The settings may change; the key principals are remarkably resilient.

But these are big topics, and it is perhaps time to return to the question of what we mean by project management and the management of projects, and to the structure of the book.

Project Management

There are several levels at which the subject of project management can be approached. We have already indicated one of them in reference to the PMI model. As we and several other of the *Guides*' authors indicate later, this is a wholly valid, but essentially delivery, or execution-oriented perspective of the discipline: what the project manager needs to do in order to deliver the project "on time, in budget, to scope." If project management professionals cannot do this effectively, they are failing at the first fence. Mastering these skills is the *sine qua non*—the 'without which nothing'—of the discipline. Volume 1 addresses this basic view of the discipline—though by no means exhaustively (there are dozens of other books on the market that do this excellently—including some outstanding textbooks: Meredith and Mantel, 2003; Gray and Larson, 2003; Pinto, 2004).

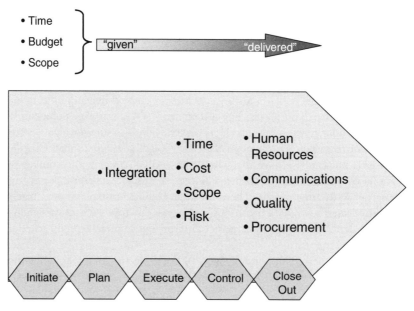

PROJECT MANAGEMENT:
"On time, in budget, to scope" execution/delivery

The overriding paradigm of project management at this level is a control one (in the cybernetic sense of control involving planning, measuring, comparing, and then adjusting performance to meet planned objectives, or adjusting the plans). Interestingly, even this model—for us, the foundation stone of the discipline—is often more than many in other disciplines think of as project management: many, for example, see it as predominantly oriented around scheduling (or even as a subset, in some management textbooks, of operations management). In fact, even in some sectors of industry, this has only recently begun to change, as can be seen towards the end of the book in the chapter on project management in the pharmaceutical industry. It is more than just scheduling of course: there is a whole range of cost, scope, quality and other control activities. But there are other important topics too.

Managing project risks, for example, is an absolutely fundamental skill even at this basic level of project management. Projects, by definition, are unique: doing the work necessary to initiate, plan, execute, control, and close-out the project will inevitably entail risks. These need to be managed.

Both these areas are mainstream and generally pretty well understood within the traditional project management community (as represented by the PMI *PMBOK*® '*Guide*' (PMI, 2004) for example). What is less well covered, perhaps, is the people-side of managing projects. Clearly people are absolutely central to effective project management; without people projects simply could not be managed. There is a huge amount of work that has been done on how organizations and people behave and perform, and much that has been written on this within a project management context (that so little of this finds its way into *PMBOK* is almost certainly due to its concentration on material that is said in *PMBOK* to be "unique" to project management). A lot of this information we have positioned in Volume 3, which deals more with the area of competencies, but some we have kept in the other volumes, deliberately to make the point that people issues are essential in project delivery.

It is thus important to provide the necessary balance to our building blocks of the discipline. For example, among the key contextual elements that set the stage for future activity is the organization's structure—so pivotal in influencing how effectively projects may be run. But organizational structure has to fit within the larger social context of the organization—its culture, values, and operating philosophy; stakeholder expectations, socioeconomic, and business context; behavioural norms, power, and informal influence processes, and so on. This takes us to our larger theme: looking at the project in its environment and managing its definition and delivery for stakeholder success: "the management of projects."

The Management of Projects

The thrust of the books is, as we have said, to expand the field of project management. This is quite deliberate. For as Morris and Hough showed in *The Anatomy of Major Projects* (1987), in a survey of the then-existing data on project overruns (drawing on over 3,600 projects as well as eight specially prepared case studies), neither poor scheduling nor even lack of teamwork figured crucially among the factors leading to the large number of unsuccessful projects in this data set. What instead were typically important were items such as client changes, poor technology management, and poor change control; changing social, economic, and environmental factors; labor issues, poor contract management, etc. Basically, the message was that while traditional project management skills are important, they are often not *sufficient* to ensure project success: what is needed is to broaden the focus to cover the management of external and front-end issues, not least technology. Similarly, at about the same time, and subsequently, Pinto and his coauthors, in their studies on project success (Pinto and Slevin, 1988; Kharbanda and Pinto, 1997), showed the importance of client issues and technology, as well as the more traditional areas of project control and people.

The result of both works has been to change the way we look at the discipline. No longer is the focus so much just on the processes and practices needed to deliver projects "to scope, in budget, on schedule," but rather on how we set up and define the project to deliver stakeholder success—on how to manage projects. In one sense, this almost makes the subject impossibly large, for now the only thing differentiating this form of management from other sorts is "the project." We need, therefore, to understand the characteristics of the project development life cycle, but also the nature of projects in organizations. This becomes the kernel of the new discipline, and there is much in this book on this.

Morris articulated this idea in *The Management of Projects* (1994, 97), and it significantly influenced the development of the Association for Project Management's Body of Knowledge as well as the International Project Management Association's Competence Baseline (Morris, 2001; Morris, Jamieson, and Shepherd, 2006; Morris, Crawford, Hodgson, Shepherd, and Thomas, 2006). As a generic term, we feel "the management of projects" still works, but it is interesting to note how the rising interest in program management and portfolio management fits comfortably into this schema. Program management is now strongly seen as the management of multiple projects connected to a shared business objective—see, for example, the chapter by Michel Thiry (*The Wiley Guide to Project, Program & Portfolio Management*, Chapter 6.) The emphasis on managing for business benefit, and on managing projects, is exactly the same as in "the management of projects." Similarly, the recently launched *Japanese Body of Knowledge, P2M (Program and Project Management)*, discussed *inter alia* in Lynn Crawford's chapter on project management standards (*The Wiley Guide to Project Organization & Project Management Competencies*, Chapter 10), is explicitly oriented around managing programs and projects to create, and optimize, business value. Systems management, strategy, value management, finance, and relations management for example are all major elements in *P2M:* few, if any, appear in *PMBOK.*

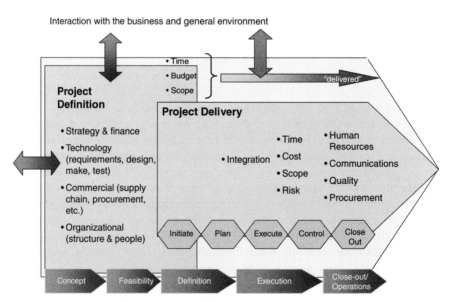

THE MANAGEMENT OF PROJECTS involves managing the definition and delivery of the project for stakeholder success. The focus is on the project in its context. Project and program management – and portfolio management, though this is less managerial – sit within this framework.

("The management of projects" model is also more relevant to the single project situation than *PMBOK* incidentally, not just because of the emphasis on value, but via the inclusion of design, technology, and definition. There are many single project management situations, such as Design & Build contracts for example, where the project management team has responsibility for elements of the project design and definition).

Structure of *The Wiley Guide to Project Control*

The Wiley Guides to the Management of Projects series is made up of four distinct, but interrelated, books:

* *The Wiley Guide to Project, Program & Portfolio Management*
* *The Wiley Guide to Project Control*
* *The Wiley Guide to Project Organization & Project Management Competencies*
* *The Wiley Guide to Project Technology, Supply Chain & Procurement Management*

This book, *The Wiley Guide to Project Control*, is based on the "traditional" project management activities—control, risk, time, cost, and quality. Project control represents more than the simple evaluation of project performance. In fact, the control cycle—as described in Chapter 1 by Pete Harpum—is typically set up around a four-stage recurring process that first challenges us to establish targets, measure performance, compare actual results with intended goals, and make necessary adjustments. While this model offers a useful conceptual backdrop, in practice we find ourselves dealing with some aspects of project management that are seemingly as old as the hills (or somewhere between fifty and 5,000 years anyway!), but also some newer concepts—like Quality Management or Risk Management. Thus, readers will find that embedded in these chapters are not only well understood concepts, but also important, cutting-edge work that has a profound impact on the manner in which many organizations are managing projects today.

1. Pete Harpum's Chapter 1 positions project control within a "systems" context, reminding us that, in the cybernetic sense, control involves planning as well as monitoring, and also taking corrective action. All the fundamental levers of project control are touched upon. But as in a proper "systems" way of looking at things, Pete reminds us that projects exist within bigger systems and hence we need to relate project control to business strategy—or the project's equivalent contextual objectives.
2. In Chapter 2, Asbjørn Rolstadås takes us, with some rigor, through the principles of project time and cost control. Gantt charts and critical path are explained together with the use of contingencies in project estimating.
3. Larry Leach, in Chapter 3, extends Asbjørn's text by offering a detailed review of critical chain project management. Beginning with Goldratt's "Theory of Constraints" (Dettmer 1997),—which, as he points out, isn't really a theory as such—Larry explains how one

of the limitations of critical path planning, resource leveling, can be treated as a constraint, and how this constraint can be exploited by more sophisticated statistical modeling of activity times. This is done through project buffering. The implications of this methodology are then reviewed—to performance measurement (see Chapter 4), resourcing, and decision-making. Multi-project critical chains and organization-wide critical chain management is then reviewed, with particular emphasis on the behavior changes that will be required by managers.

4. Reporting performance is obviously a key part of project control, but as anyone who has tried to do this soon realizes, it is not easy. The essential challenge is what measures to use, and how to report on these in an integrated way. Dan Brandon tackles this head-on in Chapter 4. Beginning, like Peter Harpum in Chapter 1, from the control cycle view, Dan covers basic schedule and cost reporting before moving on to Earned Value. He then broadens the discussion to look again at the question of project success (previously covered in Chapter 5) and how to measure this concept. As we see elsewhere in *The Wiley Guide* (George Steel for example in Vol. 4, Chapter 13), the project values and drivers need to dictate the measures that will be employed. Dan ends by discussing the Balanced Scorecard approach as a framework for project performance measurement, integrating Earned Value, and project success measures within this.

5. Managing project risks is absolutely fundamental skill at any level of management—including this "base" level of project management, as *PMBOK* recognizes. Projects by definition are unique: doing the work necessary to initiate, plan, execute, control and close-out the project inevitably entails risks. These, as Steve Simister in Chapter 5 succinctly summarizes, will need to be managed and Steve proposes a 'risk strategy-identification-analysis-response-control' process for doing so.

6. Stephen Ward and Chris Chapman, two of the leading scholars in the field of risk management, take us in Chapter 6 on a second, more critical discussion of the topic, suggesting in the process that we probably should be thinking of it now rather as Uncertainty Management than risk management. In doing so, they touch on many of the issues we have recently been discussing: estimating business benefit, design and technology risk, statistical probability, and selection of performance measures, among others.

7. Quality Management (QM) is a subject that bears both on strategic planning and operational control. Martina Hueman, in Chapter 7, traces the evolution of quality management from quality control to Total Quality Management before looking at quality management in project management. She does this under the headings of certification, accreditation, the quality "Excellence" model, benchmarking, audits and reviews, coaching and consulting, and evaluation. Quality standards for projects and programs are then discussed in detail, with reference to engineering, construction, and IT / IS. As she points out, typically much of the project management focus tends to end up being about product quality; there is however opportunity for it to be applied to project management processes and practices—and people—as she shows, not least in discussion of QM on organizational change projects. Martina's discussion is particularly valuable on the role of management audits and reviews, showing how these can lead to improvements in

project management competency. The chapter ends by noting the role of the Project Management Office, the topic of the next chapter, in supporting this process.

8. James Young and Martin Powell review the Project Management Support Office in Chapter 8. Initially, the project office was often seen as a project status reporting unit. Increasingly, however, it has become a "home" for project management and a center for project management excellence. The PMSO may be found at three levels in an organization: at the corporate/enterprise level, the business unit level, and the project/program level. It has particular applicability in organizations where there is considerable virtual working. Its activities cover portfolio, program, and project support; enterprise-wide project management support (resource planning, communications, benchmarking, performance measurement), competency development; and support in tools and techniques. James and Martin walk us through examples of how the PMSO might provide support in each of these areas. They conclude the chapter with a discussion of how PMSO effectiveness can be measured.

About the Authors

Peter Harpum

Peter Harpum is a project management consultant with INDECO Ltd, with significant experience in the training and development of senior staff. He has consulted to companies in a wide variety of industries, including retail and merchant banking, insurance, pharmaceuticals, precision engineering, rail infrastructure, and construction. Assignments range from wholesale organizational restructuring and change management, through in-depth analysis and subsequent rebuilding of program and project processes, to development of individual persons' project management capability. Peter has a deep understanding of project management processes, systems, methodologies, and the 'soft' people issues that programs and projects depend on for success. Peter has published on design management; project methodologies, control, and success factors; capability development; portfolio and program value management; and internationalization strategies of indigenous consultants. He is a Visiting Lecturer and examiner at UMIST on project management.

Stephen Simister

Dr. Simister is a consultant and lecturer in project management and a director of his own company, Oxford Management & Research Ltd., and an Associate of INDECO Ltd. His specialty is working with clients to define the scope and project requirements to meet their business needs; facilitating group decision support workshops which allow theses requirements to be articulated to outside suppliers of goods and services; and facilitating both value and risk management workshops. He has experience in most business sectors and has been involved in all stages of project lifecycles. As a Fellow of the Association for Project Management, Stephen is currently Chairman of the Contracts & Procurement Specific Interest

Group. He is also a Chartered Building Surveyor with the Royal Institution of Chartered Surveyors and sits on the construction procurement panel. Stephen lectures at a number of European universities and has written extensively on the subject of project and risk management. He is co-editor of Gower's *Handbook of Project Management*, 3rd Ed. He received his doctorate in Project Management from The University of Reading and maintains close links with the university.

Asbjørn Rolstadås

Asbjørn Rolstadås is professor of production and quality engineering at the Norwegian University of Science and Technology. His research covers topics such as numerical control of machine tools, computer-aided manufacturing systems, productivity measurement and development, computer-aided production planning and control systems, and project management methods and systems. He is a member of The Royal Norwegian Society of Sciences, The Norwegian Academy of Technical Sciences, and The Royal Swedish Academy of Engineering Sciences. He serves on the editorial board of number of journals, and is the founding editor of the International Journal of Production Planning and Control. He is past president of IFIP (International Federation for Information Processing). He is also past president of the Norwegian Computer Society. He is currently the head of the Norwegian Centre for Project Management. He has done studies of project execution of major governmental projects, mainly within development of oil and gas in the North Sea. Research on risk analyses and contingency planning in cost estimates; and developed training courses in project planning and control using e-learning.

Larry Leach

Larry Leach is the president of the Advanced Projects Institute (API), a management consulting firm. API specializes in project management, which includes leading the implementation of the new Critical Chain method of project management. Larry supports many companies, large and small, with diverse projects, ranging from R&D to construction. Larry developed and operated Project Management Office for American National Insurance Company (ANICO), in Galveston Texas, performing large IT projects. He has worked at the Vice Presidential level in several Fortune 500 companies, where he managed a variety of programs of large and small projects. Prior to that, Larry successfully managed dozens of projects, ranging up to one billion dollars. Larry has Masters' Degrees in both Business Management from the University of Idaho, and in Mechanical Engineering from the University of Connecticut. He was awarded membership in Tau Beta Pi, the Engineering honorary society, while earning his undergraduate degree in Engineering at Stevens Institute of Technology. Larry is a member of the Project Management Institute (PMI) and a certified Project Management Professional. His has published many papers on related topics, including a PMI Journal article on Critical Chain in June, 1999, and a pair of papers published in *PM Network* in spring, 2001. He presents seminars for PMI Seminars, and authored the recently published book, *Critical Chain Project Management*. Larry also serves as faculty for the University of Phoenix, facilitating courses in business management.

Daniel Brandon

Dr. Daniel Brandon is a Professor and the Department Chairperson of the Information Technology Management (ITM) Department at Christian Brothers University (CBU) in Memphis, TN. His education includes a BS in Engineering from Case Western University, MS in Engineering from the University of Connecticut, and a Ph.D. from the University of Connecticut specializing in computer control and simulation. He also holds the PMP (Project Management Professional) certification. His research interest is focused on software development, both on the technical side (analysis, design, and programming) and on the management side. In addition to his eight years at CBU, Dr. Brandon has over twenty years experience in the information systems industry including experience in general management, project management, operations, research, and development. He was the Director of Information Systems for the Prime Technical Contractor at the NASA Stennis Space Center for six years, MIS manager for Film Transit Corporation in Memphis for ten years, and affiliated with Control Data Corporation in Minneapolis for six years in several positions, including Manager of Applications Development. He is also an independent consultant and software developer for several industries including: Finance, Transportation / Logistics, Medical, Law, and Entertainment.

Stephen Ward

Stephen Ward is a Senior Lecturer in Management Science at the School of Management, University of Southampton, UK. He holds a B.Sc. in Mathematics and Physics (Nottingham), an M.Sc. in Management Science (Imperial College, London), and a PhD in developing effective models in the practice of operational research (Southampton). He is a member of the PMI and a Fellow of the UK Institute of Risk Management. Before joining Southampton University, he worked in the OR group at Nat West Bank. At Southampton, he was responsible for setting up the School's MBA program and he is now Director of the School's Master's program in Risk Management. He founded and edited for ten years the Operational Research Society's quarterly publication *OR Insight* which continues to publish articles on the application of management science. Dr Ward's teaching interests cover a wide range of management topics including decision analysis, managerial decision processes, insurance, operational and project risk management, and strategic management. His research and consulting activities relate to project risk management systems and the management of uncertainty. He has published a range of papers on risk management and co-authored two books (with Chris Chapman): *Project Risk Management* (Wiley, second edition Autumn 2003), and *Managing Project Risk and Uncertainty* (Wiley 2002). The latter text provides a case based treatment of key issues in uncertainty management using constructively simple forms of analysis. Dr. Ward is currently working on organization-wide approaches to integrated risk management, building on emergent issues in project risk management.

Chris Chapman

Chris Chapman has been Professor of Management Science, University of Southampton, since 1986. He is a former Head of the Department of Accounting and Management Science (1984–91) and Director of the School of Management (1995–98). He was founding Chair for the Project Risk Management Specific Interest Group, Association for Project Management 1986–91; President of the Operational Research Society (1992–93); and a panel member, Business and Management Studies, HEFCE Research Assessment Exercise in 1992 and 1996. He was elected Honorary Fellow of the Institute of Actuaries, 1999. His consulting and research have focused on risk management since 1975. He undertook seminal work as a consultant to BP International, developing project planning and costing procedures for their North Sea operations, 1976–82, adopted world-wide by BP. The new ideas associated with this work were developed and generalized during a variety of assignments in the USA, Canada, and the UK for many major clients. He has authored or co-authored five books, fifteen book chapters, and forty refereed academic journal papers (including ORS President's medal for 1985 paper in JORS). His most recent book is *Project Risk Management: Processes, Techniques and Insights* (with Stephen Ward), Wiley, 1997.

Martina Huemann

Dr. Martina Huemann holds a doctorate in project management of the Vienna University of Economics and Business Administration. She also studied business administration and economics at the University Lund, Sweden, and the Economic University Prague, Czech Republic. Currently, she is assistant professor in the Project Management Group of the Vienna University of Economics and Business Administration. There she teaches project management to graduate and postgraduate students. In research, she focuses on individual and organizational competences in Project-oriented Organisations and Project-oriented Societies. She is visiting fellow of The University of Technology Sydney. Martina has project management experience in organizational development, research, and marketing projects. She is certified Project Manager according to the IPMA—International Project Management Association—certification. Martina organizes the annual pm days research conference and the annual PM days student paper award to promote project management research. She contributed to the development of the PM baseline—the Austrian project management body of knowledge and is board member of Project Management Austria—the Austrian project management association. She is assessor of the IPMA Award and trainer of the IPMA advanced courses. Martina is trainer and consultant of Roland Gareis Consulting. She has experience with project-oriented organizations of different industries and the public sector. Martina is specialized in management audits and reviews of projects and programs, and human resource management issues like project management assessment centers for project and program managers.

James Young

James Young is a Senior Consultant with INDECO Ltd. He is highly experienced in the development of organizational and individual project management competencies, as well as advising companies on strategies for the successful delivery of portfolios, programs, and projects. He has undertaken a number of high-profile initiatives for major blue-chip organizations. He brings experience from work undertaken across Europe, Scandinavia, South America, and the United States. Earlier in his career, James worked in the UK construction industry. James has authored a number of published articles on various aspects of project management. His research has been presented at the World Congress for Project Management, the Project Management Institute's European Conferences and the International Project Managers Association Conference. James has a first class degree from UMIST.

Martin Powell

Martin Powell is a senior consultant at INDECO International Management Consultants. He is responsible for the delivery of assignments, in both the pharmaceutical and oil and gas sectors. He has been working with a major pharmaceutical company developing a global implementation strategy for a project management support office as well as providing strategic advice to a number of companies looking to develop one. He leads teams in the development of tailored guidelines as well as supporting wider initiatives in knowledge management and communications. Prior to joining INDECO, Martin worked as a Project Engineer for Impresa Federici S.p.a, an Italian Engineering Company in Rome, and as a Project Manager for Ove Arup & Partners, an engineering design consultancy, where he was responsible for the delivery of several high-profile projects in Italy, Spain and, Asia. He also worked with many of the leading architects such as Richard Rogers, Renzo Piano, and Zaha Hadid, supporting various Master Planning submissions for design competitions as well as being involved in troubleshooting projects. Martin has a degree in Civil Engineering from the University of Dundee. He also took electives in Spanish at St. Andrews University. He also obtained a scholarship to study at Stevens Institiute of Technology in Hoboken, New Jersey.

The Wiley Guides to the Management of Projects series offers an opportunity to take a step back and evaluate the status of the field, particularly in terms of scholarship and intellectual contributions, some twenty-four years after Cleland and King's seminal *Handbook*. Much has changed in the interim. The discipline has broadened considerably—where once projects were the primary focus of a few industries, today they are literally the dominant way of organizing business in sectors as diverse as insurance and manufacturing, software engineering and utilities. But as projects have been recognized as primary, critical organizational forms, so has recognition that the range of practices, processes, and issues needed to manage them is substantially broader than was typically seen nearly a quarter of a century ago. The old project management "initiate, plan, execute, control, and close" model once considered the basis for the discipline is now increasingly recognized as insufficient and inadequate, as the many chapters of this book surely demonstrate.

The shift from "project management" to "the management of projects" is no mere linguistic sleight-of-hand: it represents a profound change in the manner in which we approach projects, organize, perform, and evaluate them.

On a personal note, we, the editors, have been both gratified and humbled by the willingness of the authors (very busy people all) to commit their time and labor to this project (and our thanks too to Gill Hypher for all her administrative assistance). Asking an internationally recognized set of experts to provide leading edge work in their respective fields, while ensuring that it is equally useful for scholars and practitioners alike, is a formidable challenge. The contributors rose to meet this challenge wonderfully, as we are sure you, our readers, will agree. In many ways, the *Wiley Guides* represent not only the current state of the art in the discipline; it also showcases the talents and insights of the field's top scholars, thinkers, practitioners, and consultants.

Cleland and King's original *Project Management Handbook* spawned many imitators; we hope with this book that it has acquired a worthy successor.

References

Christensen, C. M. 1999. *Innovation and the General Manager.* Boston: Irwin McGraw-Hill.

Cleland, D. I., and King, W. R. 1983. *Project Management Handbook.* New York: Van Nostrand Reinhold.

Cleland, D. I. 1990. *Project Management: Strategic Design and Implementation.* Blue Ridge Summit, PA: TAB Books.

Gray, C. F., and E. W. Larson. 2003. *Project Management.* Burr Ridge, IL: McGraw-Hill.

Griseri, P. 2002. *Management Knowledge: a critical view.* London: Palgrave.

Kharbanda, O. P., and J. K. Pinto. 1997. *What Made Gertie Gallop?* New York: Van Nostrand Reinhold.

Meredith, J. R. and S. J. Mantel, *Project Management: A Managerial Approach.* 5th Edition. New York: Wiley.

Morris, P. W. G., and G. H. Hough. 1987. *The Anatomy of Major Projects.* Chichester: John Wiley & Sons Ltd.

Morris, P. W. G. 1994. *The Management of Projects.* London: Thomas Telford; distributed in the USA by The American Society of Civil Engineers; paperback edition 1997.

Morris, P. W. G. 2001. "Updating the Project Management Bodies Of Knowledge" *Project Management Journal* 32(3):21–30.

Morris, P. W. G., H. A. J. Jamieson, and M. M. Shepherd. 2006. "Research updating the APM Body of Knowledge 4th edition" *International Journal of Project Management* 24:461–473.

Morris, P. W. G., L. Crawford, D. Hodgson, M. M. Shepherd, and J. Thomas. 2006. "Exploring the Role of Formal Bodies of Knowledge in Defining a Profession—the case of Project Management" *International Journal of Project Management* 24:710–721.

Pinto, J. K. and D. P. Slevin. 1988. "Project success: definitions and measurement techniques," *Project Management Journal* 19(1):67–72.

Pinto, J. K. 2004. *Project Management.* Upper Saddle River, NJ: Prentice-Hall.

Project Management Institute. 2004. *Guide to the Project Management Body of Knowledge*. Newtown Square, PA: PMI.

Project Management Institute. 2003. *Organizational Project Management Maturity Model*. Newtown Square, PA: PMI.

Wheelwright, S. C. & Clark, K. B. 1992. *Revolutionizing New Product Development*. New York: The Free Press.

THE WILEY GUIDE TO PROJECT CONTROL

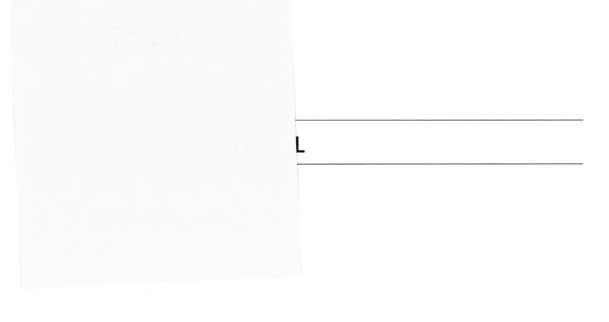

L

Project control is about ensuring that the project delivers what it is set up to deliver. Fundamentally, the process of project control deals with ensuring that other project processes are operating properly. It is these other processes that will deliver the project's products, which in turn will create the change desired by the project's sponsor. This chapter provides an overview of the project control processes, in order to provide the conceptual framework for the rest of this section of the book.

Introduction

Control is fundamental to all management endeavor. To manage implies that control must be exercised. Peter Checkland connects the two concepts as follows:

> The management process. . .is concerned with deciding to do or not to do something, with planning, with alternatives, with monitoring performance, with collaborating with other people or achieving ends through others; it is the process of taking decisions in social systems in the face of problems which may not be self generated.
>
> Checkland, 1981

In short to

- plan
- monitor
- take action

One may ask what is the difference between project control and any other type of management control? Fundamentally there is little that project managers must do to control their work that a line manager does not do. Managers of lines and projects are both concerned with planning work; ensuring it is carried out effectively (the output from the work "does the job") and efficiently (the work is carried out at minimum effort and cost). Ultimately, managers of lines and projects are concerned with delivering what the customer wants. The line management function is usually focussed on maximizing the efficiency of an existing set of processes—by gradual and incremental change—for as long as the processes are needed. The objective of operations management (or "business–as–usual") is rarely to create change of significant magnitude. Projects, on the other hand, are trying to reach a predefined end state that is different to the state of affairs currently existing; projects exist to create change. Because of this, projects are almost always time-bound. Hence, the significant difference is not in control per se, but in the processes that are being controlled—and in the focus of that control.

Project management is seen by many people as mechanistic (rigidly follow set processes and controlled by specialist tools, apropos a machine) in its approach. This is unsurprising given that the modern origins of the profession lie in the hard-nosed world of defense industry contracting in America. These defense projects (for example, the Atlas and Polaris missiles) were essentially very large systems engineering programs where it was important to schedule work in the most efficient manner possible. Most of the main scheduling tools had been invented by the mid-1960s. In fact, virtually all the mainstream project control techniques were in use by the late 1960s. A host of other project control tools were all available to the project manager by the 1970s, such as resource management, work breakdown structures, risk management, earned value, quality engineering, configuration management, and systems analysis (Morris, 1997).

The reality, of course, is that project management has another, equally important aspect to it. Since the beginning of the 1970s research has shown that project success is not dependent only on the effective use of these mechanistic tools. Those elements of project management to do with managing people and the project's environment (leadership, team building, negotiation, motivation, stakeholder management, and others) have been shown to have a huge impact on the success, or otherwise, of projects (Morris, 1987; Pinto and Slevin, 1987—see also Chapter 5 by Brandon. Both these two aspects of project management—"mechanistic" control and "soft," people-orientated skills—are of equal importance, and this chapter does not set out to put project control in a position of dominance in the project management process. Nevertheless, it is clear that effective control of the resources available to the project manager (time, money, people, equipment) is central to delivering change. This chapter explains why effective control is fundamentally a requirement for project success.

The first part of the chapter explains the concept of control, starting with a brief outline of systems theory and how it is applied in practice to project control. The second part of the chapter outlines the project planning process—before project work can be controlled, it is critical that the work to be carried out is defined. Finally, the chapter brings project planning and control together, describing how variance from the plan is identified using performance measurement techniques.

Project Control and Systems Theory

Underlying control theory in the management sciences is the concept of the *system*. The way that a project is controlled is fundamentally based on the concept of system control—in this case the system represents the project. Taking a systems approach leads to an understanding of how projects function in the environment in which they exist. A system describes, in a holistic manner, how groups are related to each other. These may, for instance, be groups of people, groups of technical equipment, groups of procedures, and so on. (In fact, the systems approach grew out of general systems theory, which sought to understand the concept of "wholeness"—see Bertalanffy, 1969.) The systems approach exists within the same conceptual framework as a project; namely, to facilitate change from an initial starting position to an identified final position.

The Basic Open System

A closed system is primarily differentiated from an open system in that the former has impermeable boundaries to the environment (what goes on outside the system does not affect the system), while the latter has permeable boundaries (the environment can penetrate the boundary and therefore affect the system). In a closed system, fixed "laws" enable accurate predictions of future events to be made. A typical example of this is in physical systems (say, for instance, a lever) where a known and unchanging equation can be used to predict exactly what the result will be of applying a force to one part of the system. Open systems do not allow such accurate predictions to be made about the future, because many influences cross the boundary and interact with the system, making the creation of predictive laws impossible.

The key feature of the open system approach that makes it useful in the analysis and control of change is that the theory demands a holistic approach be taken to understanding the processes and the context they are embedded in. It ensures that account is taken of all relevant factors, inputs or influences on the system, and its environmental context. Another key feature of an open system is that the boundaries are to a large extent set arbitrarily, depending on the observer's perspective. Moreover, wherever the boundaries are placed, they are still always permeable to energy and information from the outside. It is this quality that allows the relationships between the system and its environment to be considered in the context of change, from an initial condition to a final one.

A critical part of this theory that is useful when considering management control is that the open system always moves toward the achievement of superordinate goals. This means that although there may be conflicting immediate goals within the creative transformation process(es), the overall system moves toward predefined goals that benefit the system as a whole (see Katz and Kahn, 1969). In a project system these goals are the project objectives.

In simple terms the basic open system model is shown in Figure 1.1.

The Open System Model Applied to Project Control

If the generic system diagram is redrawn to represent a project and its environment, the relationship of control to planning (how the project is going to achieve its objectives) can be made clear (see Figure 1.2).

FIGURE 1.1. THE BASIC SYSTEMS MODEL.

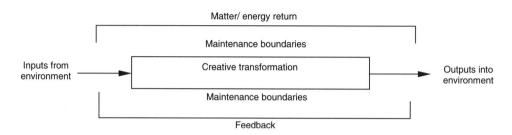

The system: A set of components that are interrelated, acting in a unified way, to achieve a goal.

The creative transformation: The process(es) that act on the energy, resources, and materials from the environment to turn inputs into outputs.

Inputs: Energy, resources, and materials from the environment.

Outputs: Products, knowledge, or services that help the system achieve its goals.

Maintenance boundaries: The mechanism that defines the transformation process(es)—and therefore creates the system's unique identity.

Matter/ energy return: Matter and energy are returned to the environment in order to ensure the system remains in equilibrium.

Feedback: The feedback required to ensure the system stays on course to deliver its goals.

The environment: These are all the influences that act on the system, and the control system attempts to mitigate.

Source: After Jackson (1993).

The value of the open system model of the project shows how the "mechanistic" control meta process attempts to ensure that the project continues toward achievement of its objectives and the overall goal. The "softer" behavioral issues are evident throughout the project system: within the project processes, acting across the permeable boundaries with the project stakeholders and wider environment, and indeed around the "outside" of the project system but nevertheless affecting the project goal—and hence the direction the project needs to move in to reach that goal.

The boundaries of the project are defined by the project plans. These plans define what the project processes need to do to reach the system's goals (defined by the project's environment). The plans also determine *where* in the organizational hierarchy the project exists, because projects have subsystems (work packages) and exist within a supra-system (programs and portfolios of projects). This is shown in Figure 1.3.

FIGURE 1.2. THE PROJECT AS A SYSTEM.

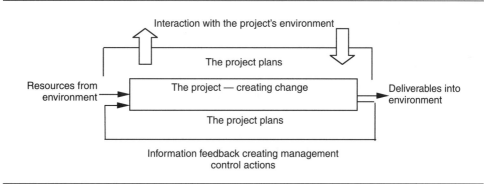

FIGURE 1.3. THE HIERARCHICAL NATURE OF PROJECT CONTROL SYSTEMS.

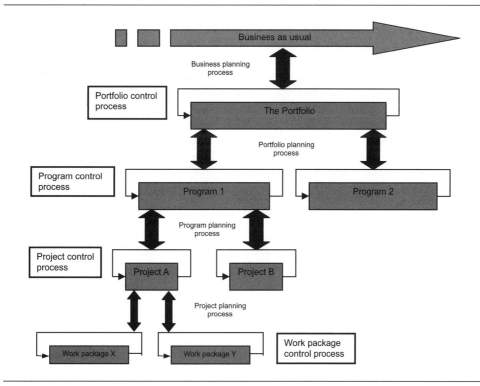

To make the distinctions clear between portfolio and program, their definitions are listed in the following, elaborated for clarity from the Association for Project Management Body of Knowledge, 4th edition (APM, 2000):

Program management	Often a series of projects are required to implement strategic change. Controlling a series of projects that are directed toward a related goal is program management. The program seeks to integrate the business strategy, or part of it, and the projects that will implement that strategy in an overarching framework.
Portfolio management	In contrast to a program, a portfolio comprises a number of projects, or programs, that are not necessarily linked by common objectives (other than at the highest level), but rather are grouped together to enable better control to be exercised over them.

The feedback loop measures where the project is deviating from its route (the plans) to achieving the project goals and provides inputs to the system to correct the deviation. Control is therefore central to the project system; it tries to ensure that the project stays on course to meet its objectives and to fulfill its goals. The deviation away from the project's goals can be caused by suboptimal project processes (poor plan definition, for instance) or by positive or negative influences from the environment penetrating the permeable boundaries and affecting the processes or goal (poor productivity, failures of technologies to perform as expected, market changes, political influence, project goals being changed, and a host of similar inputs).

At each of these levels of management there is a planning process. This process ensures alignment between objectives of work at different levels, for example, between programs and projects—the program plan. Consequently, the control of these systems is hierarchical in nature.

The abstract models described so far can be used to diagrammatically show the overarching project control process, in which all the system elements are combined. This process is shown in Figure 1.4. In this diagram the way in which the project life cycle stages of initiate (define objectives), plan, and implement (carry out work) are overlaid with the control process is clearly shown. The next part of this chapter describes how the project plan is developed; that is, how the project system is defined.

Defining the Project Objectives

The clear and unambiguous definition of project objectives is fundamental to achieving project success. However, prior to project definition it is necessary to understand the business strategy, or at least that part of the strategy, being delivered (or facilitated) by the project. If the strategic goal is not understood, there is little chance of the project's objectives being accurately defined.

There are various definitions of strategy, viz:

FIGURE 1.4. THE PROJECT CONTROL PROCESS.

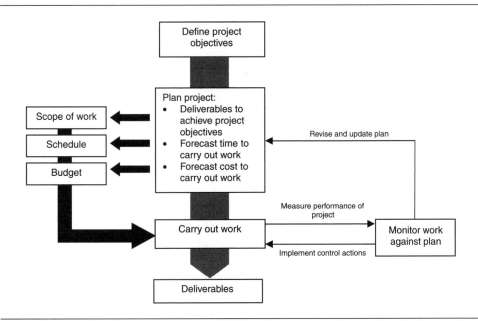

- "Strategic thinking is the art of outdoing an adversary, knowing that the adversary is trying to do the same to you" (Dixit and Nalebuff, 1991).
- ". . .the general direction in which the [company's] objectives are to be pursued" (Cleland and King, 1983).
- " . . . strategies embrace those patterns of high leverage decisions (on major goals, policies, and action sequences) which affect the viability and direction of the entire enterprise or determine its competitive posture for an extended period of time" (Quinn, 1978).

Business strategies have dual functions; firstly to communicate the strategy at a detailed level and identify the method of implementation (in part by programs and projects), and secondly to act as a control device. Both of these functions rely on the strategy having the characteristic of a plan—in other words, strategy is represented in a decomposed and articulated form. The communication aspect of the program informs people in the organization (and those external to it) of the intended strategy and the consequences of the strategy being implemented. They not only communicate the intention of the strategy but also the role that the employees have to take in its implementation (project, nonproject, or business-as-usual work). The control aspect of the strategic program assesses not only performance toward the implementation of strategy but also behavior of the organization as the strategic actions take effect—has the behavior of the firm adjusted as predicted by the strategy? This then forms the feedback loop anticipated in the control system. See Figure 1.5.

FIGURE 1.5. MECHANISM FOR ACHIEVING STRATEGIC CHANGE.

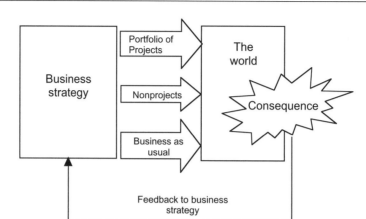

Business strategy provides two or three high-level project objectives, and from these are developed additional project specific objectives and the *project* strategy. Traditionally, project objectives have been defined in terms of the "project triumvirate" of time to complete, cost to complete, and adherence to technical specifications (i.e., quality) (Barnes, 1988). This does not mean that other objectives should not be considered. Objectives for a project to build an oil platform, for example, could be stated as follows:

Primary objectives

- *Safety*. Minimum number of accidents
- *Operability*. Minimum number of days downtime
- *Time*. Maximum acceptable duration before start-up
- *Cost*. Through-life cost for maximum business benefit

Secondary objectives

- *Reliability*. Minimum number of failures per month
- *Ease of installation*. Non-weather-dependent process
- *Maintainability*. Minimum number of maintenance staff required

It is important to reduce uncertainty to the minimum for a project, and setting clear and prioritized objectives is a fundamental part of this process. However, sometimes changes to the objectives become inevitable. Occasionally the environment changes unexpectedly—for example, new legislation may be introduced; economic conditions may change; business conditions may alter. That this may happen is not necessarily in itself a bad thing, or a failure of either the sponsoring organization's management or project management. Such changes may impact on the organization, and its projects, to such an extent that organizational strategy has to be changed and projects either canceled or their objectives changed to meet the needs of the new strategy.

Planning the Project

The essence of project planning is determining what needs to be created to deliver the project objectives (the project deliverables or products), and within what constraints (of time, cost, and quality). Although this may seem like a statement of the obvious, many projects still fail to meet some or all of their objectives because of inadequate definition of the work required to achieve those objectives. Planning must also consider multiple other factors in the project's environment if it is to have any real chance of success—the critical success factors discussed later in the chapter. There are a number of processes that need to be followed to plan projects effectively (see Project Management Institute, 2000):

- Define the deliverables
- Define the work packages
- Estimate the work
- Schedule the work packages
- Manage resource availability
- Create the budget
- Integrate schedule and budget
- Identify key performance indicators
- Identify critical success factors

Each of these processes are briefly described in this section of the chapter (and are described in detail in subsequent chapters of the book).

Defining the Deliverables

Projects are run to create change, and the change is defined by the objectives set for the project. The way in which objectives are achieved is by organizing work (the creative transformation from the basic systems model) to deliver tangible and intangible products into the environment that is to be changed. Therefore, it is central to project success that the specific set of deliverables required is understood and articulated. This set of deliverables forms the project scope.

Accurately defining the project scope entails five subprocesses. Each of these process steps can be highly specialist in nature for projects delivering sophisticated products or services. For this overview chapter they are described briefly in the following paragraphs.

The first of these subprocesses is *requirements definition*—understanding what is required to create the change required from running the project. Requirements are "needs" to be satisfied; they are not the solutions to deliver the change (Eisner, 1997). They are the essential starting point for determining what deliverables need to be made by the project.

Poor requirements definition and management has been found to be one of the primary contributory factors leading to project failure. There is little point in managing a project perfectly if the project's deliverables do not solve the right problem, or provide the necessary capability to the organization. Requirements definition consists of the following elements:

Gathering the project requirements	This is partly art and partly science (considered by many to be more art usually), particularly when seeking to draw out from the project stakeholders and document as complete a set of desired requirements as possible.
Assessing the requirements	The analysis and definition of business and technical requirements to assess the

- project's and organization's technical capability to deliver them
- priorities of the project's requirements, taking into consideration
  —the perceived importance of each requirement to create the change needed
  —the availability of resource (time, people, money, materials) to deliver the requirements
  —the technical capability to deliver the requirements (the requirements may be unachievable technically)
  —the risk profile that the project is able to manage effectively

	It is often necessary to iterate the assessment to get a set of requirements that will deliver the entire change desired.
Creating an adequate testing regime	In order to be sure that all the conditions to create the change have been met, it must be possible to test that the requirements have been satisfied.

In order for requirements statements to be used efficiently they should be

Structured	The project requirements should be clearly linked to the need to create change, and this is done through matching requirements to objectives.
Traceable	It should be possible to identify the source of each requirement and trace any changes to the requirements definition, and of the emerging solution to the requirements, as the project evolves.
Testable	There should be clear acceptance criteria for each requirement.

After defining the requirements a number of *conceptual designs* are created, the options for delivering the change. This process is highly creative and seeks to find efficient solutions to meet the requirements. Whenever solutions are being sought, there are always trade-offs to be considered. Each solution will have with it a set of constraints in terms of what resource is needed to create the solution; that is to say, each solution will have different needs for money, people, time, and materials. The point of generating a number of solutions is to enable decisions to be taken on what is the most effective trade-off to make for satisfying the project requirements and hence delivering the change required of the project. At this point the *concept design decision gate* is reached—the various concept design options are analyzed in the context of the change that is required to be delivered by the project. There will usually be an economic analysis to determine the following:

- Financial viability of each option
- Schedule to deliver the solution
- Technical capability of the project organization to create the solution
- Availability of suitable materials to create the solution

When the decision is made, it is important that the complete set of deliverables defined by the concept design is clearly documented.

Once the concept design has been selected, the deliverables that form that design must be *specified*—the exact details of the particular set of deliverables must be established. This is obviously important for those carrying out the work to make the deliverables. It is also fundamental to the control process (Reinertsen, 1997). It is against this specification that the project deliverables will be measured; have the deliverables created by the project been made as specified? (This is part of the quality management process). As with the previous subprocesses involved in defining the scope, specifying deliverables *can* be a complex and sophisticated task. There are essentially two ways to specify a deliverable:

Performance specification	This type of specification is stated in terms of required results, with criteria for verifying compliance—without stating methods for achieving the required results. (At a work package level the performance specification defines the functional requirements for the deliverable, the environment in which it must operate, and the interface and interchangeability requirements.)
Detailed specification	The opposite of a performance specification is a detail specification. A detail specification gives design solutions, such as how a requirement is to be achieved or how an item is to be fabricated or constructed.

After the project scope has been defined as described to this point, a final process to manage change to the scope must be established. *Scope change control* is a critical part of the overall project control meta process. Projects frequently suffer from poor scope change control, leading to the wrong deliverables being produced by the project, which means failing to satisfy the project requirements and ultimately, of course, not delivering the change that the project was set up to create. For this reason, change control is considered one of the "iron rules" of effective project management.

It is also important to realize that scope change is sometimes inevitable within the life cycle of a project. Defining requirements is dependent on having information available on what change the project is set up to achieve. It is rare that all this information is available at the beginning of the project; more usually, as the project scope is developed, additional information becomes available on the true nature of the project requirements, which means that changes to the scope are required.

The management of this scope change needs to be thorough and strictly controlled (Project Management Institute, 2000; Dixon, 2000). The change control process incorporates the following elements:

Identifying changes to scope	What new or changed deliverables are required to meet the newly identified, or more clearly understood, requirement. It also includes transmitting the request for scope change to the project's management.
Assessing the need for the scope change	This includes deciding whether the change requested is genuinely needed to meet the requirements, any implications on the entire set of project deliverables, and the impact on project constraints (time, money, people, material).
Accepting or rejecting the scope change request	This includes documenting the reasons for the decision and communicating the changed set of deliverables (or that part of the deliverables changed) to those making them and to other project stakeholders.
Adjusting the project plan	This is done to take account of the changed set of deliverables (meaning changes to budget, schedule, people carrying out the work, etc.).

FIGURE 1.6. EASE OF CHANGE COMPARED TO COST OF CHANGE OVER THE PROJECT LIFE CYCLE.

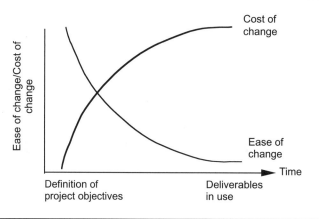

Source: Developed from Allinson (1997).

Figure 1.6 shows how the cost of changes on the project increases dramatically once the project has entered the implementation stages, compared with the much lower cost of change during the concept, feasibility, and design stages. During the early stages, fewer people are involved and the decisions made are more strategic in nature. A simple example is a change fed back from the corporate executive, at the concept stage of an organizational change project, to have separate sales and marketing departments instead of a combined one. This requires reworking the project objectives and reassessing the risk associated with the change on the overall project. It can be carried out by a small number of people relatively quickly. This same change, brought into the project during the implementation stages, will require significant amounts of time and resource to adjust the project plan to meet the new requirement. It may also cause demotivation in the project team, as work already implemented has to be "undone" and the new structure put in place.

It is worth remembering that objectives may need to be changed during the project, reflecting the reality that situations change over time. If this is the case, it may be decided that the best course of action is to complete the project (because some of its objectives are still valid and/or the cost of cancellation would outweigh the benefits of continuing) but accept a lower effectiveness of the deliverables.

A number of specialist project management techniques can be used to help in the scope definition and change control processes and are described in detail in other chapters of this book. They include the following:

Configuration management	The definition and control of how all the deliverables are configured; how they all "fit" together
Interface control	The exact specification of the interfaces between different deliverables
Systems engineering	The way in which a set of deliverables are arranged within a hierarchical "systems architecture"

(The chapter by Cooper and Reichelt addresses the issue of managing changes in more detail.)

Defining the Work Packages

Three tools are used to define the work packages:

- *Product breakdown structure (PBS)*. What needs to be made by the project.
- *Work breakdown structure (WBS)*. The work required to make these products
- *Organizational breakdown structure (OBS)*. Where in the organization the skills reside for doing the work needed

The first part of the work package definition process is to break down the main set of deliverables (identified in the scope process) into their component parts—the deliverables breakdown structure, more commonly known as the *product breakdown structure* (PBS). The disaggregation of the deliverables is developed to the level of detail that is needed by those working on the project, and commensurate with the degree of control that is required to be exercised. The work associated with making the deliverables is also divided into discrete work packages—and documented in the *work breakdown structure* (WBS). The two models must be consistent with each other; the work packages identified in the WBS must be associated with specific deliverables (i.e., products). The PBS and WBS are often combined together, and when this is the case, the diagram is usually called the WBS.

The decomposition of deliverables, and associated work, is fed into the processes for creating the forecasts of time and cost to make them; this is the estimating process. Without a clear understanding of the finite elements that need to be made by the project, it would be very difficult to carry out effective estimation of the duration to complete the tasks required and the cost to make the deliverables.

Fundamental to the planning process is deciding who will be carrying out the project work, documenting this information, and communicating it to the project team. The allocation of people to work packages is recorded in the *organizational breakdown structure* (OBS). The human resources needed to undertake the tasks to make the deliverables are often in short supply. This means that there will rarely be enough suitably skilled people available to create the deliverables as quickly as may be desired. The resources available for the work will ultimately determine the time to make the deliverables.

Estimating the Work

Forecasting how long a work package will take to complete and the cost to carry out that work is essential to effective planning. There are a number of techniques used to estimate time and cost. Essentially, the estimating process is iterative. A number of estimates are produced, reviewed, validated against the availability of resources required for the work packages, and revised accordingly.

In the estimating process, it is important to refer to historical information on the cost and time taken to carry out the same or similar work packages. Since cost is normally directly related to time (because time to complete work packages is mainly dependent on people and materials), time estimates are produced first. The cost estimate is then generated based on the forecast time to complete the work packages and the cost of materials needed:

Time estimating	Time estimates are developed by calculating how long the work package will take to complete. The inputs for the estimates of duration typically originate from the person or group on the project team who is most familiar with the nature of the tasks required to complete the work package.
Cost estimating	Cost estimating involves calculating the costs of the resources needed to complete project activities. This means the cost of peoples' time must be known, as well as the cost of materials needed to make the deliverables. This includes identifying the project management overhead—the cost of managing the project.

With estimating, there is uncertainty about the exact duration and exact cost of a work package—by definition. The uncertainty can be reflected by estimating the range within which the duration and cost for each work package will fall. The optimistic, most likely, and pessimist values for each can be provided—known as *three-point estimating*. This information can then be fed into the scheduling and budgeting processes to provide a more realistic view of the ranges of outcomes for the project as a whole. (A number of different probability distribution curves can be used.)

A fourth tool used in project planning can now be used—the *cost breakdown structure* (CBS). This documents the cost to carry out each work package, taken from the cost estimate. The information is set out in an integrated way with the PBS, WBS, and OBS to form the fundamental framework for project control. These four fundamental tools describe what has to be made to meet the project requirements, what work is needed to be carried out to make the deliverables, who is allocated to the work packages, and what the cost to create the set of deliverables will be. In large and complex projects, this information can be combined into a three-dimensional matrix (called the *cost cube*) to show the cost per product or deliverable, per resource (Turner and Remenyi, 1995). See Figure 1.7.

One advantage of creating the cost cube is that it provides a framework, or structure, for developing the estimate. For instance, one can more easily identify whether all the appropriate elements of cost associated with resources for a particular product have been included. It also enables the summing along any plane of "cubes" to provide cost information for any of the discrete items on the three axes. For instance, it is a straightforward matter to identify the total cost of resource R4 to the project by adding together the costs in each cube of the plane, as shown on the diagram.

FIGURE 1.7. THE COST CUBE MATRIX.

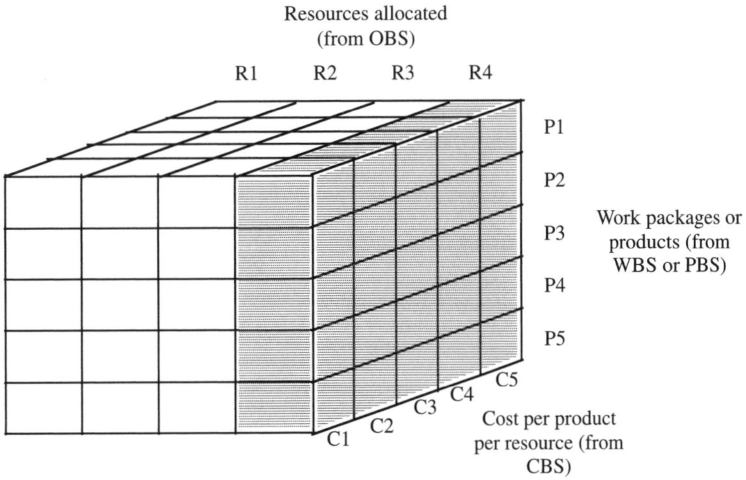

Source: Adapted from Turner and Remenyi (1995).

Scheduling the Work Packages

The essence of scheduling work packages is simple. The following factors must be known:

- Upon what previous work each subsequent work package depends
- The estimated duration of the work package
- How much "float" is available for the work—whether the work package must be carried out within a certain time or whether the time period in which it must be done can float between two known extremes

It is the combination of this information that determines what can be done when.

There are a number of well-known and commonly used techniques for modeling project work (critical path method, program evaluation and review technique, precedence diagramming, amongst others). All these techniques aim to provide flexibility in the manipulation of the model information until the optimum solution is found to suit the particular work packages of the project. Some were invented in the field of operations research, whilst others were developed by organizations for their own use. These modeling techniques are able to create projects schedules, either directly or by feeding into other techniques.

There are two distinct approaches to these scheduling techniques: *activity-on-the-arrow*, depicted in Figure 1.8, and *activity-on-the-node*, depicted in Figure 1.9. They both model the sequence of work packages by using nodes and arrows to build up a diagram that shows dependency and time for all the work packages for the project. It is then a relatively straightforward (and using PC-based software, quick) task to calculate how long it will take to complete all the work packages and, hence, the project's overall schedule.

The route through the network that determines the shortest possible time to complete all the work packages is called the *critical path*. This is important information for the project manager because these work packages will be the focus of management attention, particularly those projects for which completion "on schedule" is critical.

FIGURE 1.8. ACTIVITY-ON-THE-ARROW NETWORK.

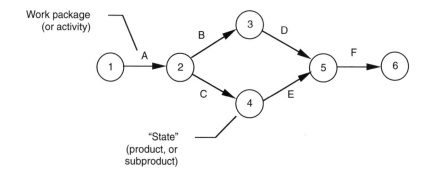

FIGURE 1.9. ACTIVITY-ON-THE-NODE NETWORK.

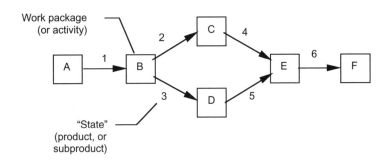

The differences between the two basic network modeling techniques appear trivial at first sight of the diagrams. However, the two approaches have significant differences in the operation of the logic used. Both have advantages and disadvantages.

Once the schedule has been established from the network diagram, it is often presented graphically on a Gantt chart. This format makes it easier to see when the work packages will be carried out in relation to each other. It also allows simple graphical representation of work completed at a given point in time, making reporting of project progress easier to show. Many current software-based scheduling tools allow the user to enter time duration for work packages directly into a Gantt chart, without first going through the process of building a network diagram. This is user-friendly but does not necessarily lead to more effective scheduling!

The schedule must be reassessed in light of the availability of people (and indeed materials—particularly those being supplied by third parties and contractors to the core project team). This part of the scheduling process is called *resource leveling*—making the schedule fit the available resources. There is likely to be an impact on the cost to complete the project, so the budget must also be reassessed (hence, the estimate is progressively elaborated and becomes progressively more accurate). Understanding the causes of variance between the actual time taken to complete the work packages and the schedule is important information for the estimating process in future projects. Similarly, variances between the actual cost to carry out the work packages and the estimated cost will also provide valuable historical estimating information.

The risk management and estimating processes will identify where there is uncertainty in the project. This uncertainty can be modeled in the schedule by allowing extra time for the work packages likely to be affected. It is clear that the entire planning process is iterative, and a number of cycles of scheduling, budgeting, and assessment of resource availability and productivity are required before a final project plan can be established.

Managing Resource Availability

The initial estimates of time and cost to complete the project are ideal estimates; the assumptions are made that sufficient people, materials, facilities, equipment, and services will be available to carry out the project at the maximum efficiency. However, before the schedule and budget can be finalized, the impact of resource availability, and the productivity of those resources, must be taken into account. For example:

- Sufficient people are rarely available (particularly where a few experts must input to many work packages).
- "Ideal" materials are often either not available where and when required, or their cost would make the project untenable (meaning less efficient materials need to be used).
- Equipment is often expensive to use and therefore must be shared across a number of projects.
- The same applies for facilities and services.

In addition to this, the resource "profile" (the types of resources needed at different times in the project) usually changes over the project life cycle. The process for managing resource allocation can be broken down into five stages:

Planning resource allocation	Identifying the types of resources required, based on the information defined in the PBS, WBS, OBS, and CBS.
Allocating resources	Coordinating the availability of resources with the suppliers of those resources and allocating them to work packages: be they internal to the organization within which the project exists (and this is commonly a major task for people—human resources—where organizations have a matrix structure) or external, such as third-party suppliers of materials, equipment, services, and facilities.
Optimizing the schedule	Inputting resource availability in the schedule, which normally means having to use the technique of resource leveling—"smoothing" resource usage to balance schedule and the availability of resources.
Monitoring resource allocation	Tracking resource usage and identifying and resolving conflicts associated with resource availability as this, and the project's needs, change over time.
Reviewing and revizing the resource allocation	Modeling the impact of changing resource use and availability on the project budget and schedule

Productivity of resources clearly has a significant influence on the schedule, and hence the cost, of the project. Productivity of equipment is often fairly easy to measure and predict; predicting productivity of people is a far more complex thing to do (and predicting productivity of highly creative design resources even more difficult). Productivity information can be gained from historical records of performance on similar work—for people and equipment. Finding and using this information is vital to effective resource allocation and resource "smoothing" of schedules.

Budgeting

The costs to complete all the work packages are identified in the estimating process. Combining the cost information with the schedule allows the cash flow curve to be created. This curve is a key piece of control information. This is particularly the case for organizations that are contracted to deliver projects on behalf of a client. These types of projects have large cash outflows (to pay for material and human resources), which are usually then charged on to the client sometime after the expenditure is incurred. If this is not managed very carefully, the project can become heavily indebted. The difference between what has been spent and what has been recovered by a project, at a given time, can cause the funders of this difference (the owners of the firm running the project) to become insolvent. Hence, effective cash flow management is a highly valued skill in project-based industries, such as construction and engineering, where huge amounts of money flow through the project.

The cash flow curve and the cost forecast are the basis of the project budget. This describes what amounts of money will be spent, on what resources, and when they will be spent. Before the cost forecast becomes a budget (the budget is the *agreed* amount of money that the project manager can spend), the effect of risk on the project needs to be assessed in cost terms and then added to the forecast. This additional amount of money allows the project manager to deal with "certain uncertainty" in the forecasts (the uncertainty can be predicted through the risk management process), and also "uncertain uncertainty" that can affect the project (uncertainty that cannot be assessed—as an example, a key human resource may leave the project without warning). These extra sums of money are the budget contingencies.

The importance of creating accurate budgets, and controlling against the budget, is obvious in a commercial environment—overspending reduces profit; underspending (whilst still making the correct project deliverables) increases profit. In the not-for-profit sector, control of money is clearly still critical. The budget document therefore identifies, line by line (hence the term "line item"), how much money is agreed to be spent per deliverable, or part deliverable. It is this detailed breakdown against which actual costs are measured and reported, and control action initiated.

After the forecasts have been analyzed and the deliverables set has been revisited to seek optimization of all the project constraints, a schedule and budget are agreed upon by the project sponsors. The budget and schedule are absolutely fundamental to the control process. It is against these two documents that the progress of the project in carrying out the work packages, and hence the production of the deliverables, is measured (see Figure 1.4).

However, having two separate documents means there exists a lack of *integrated* information related to deliverables (or products), schedule, and the budget to produce those deliverables—the complete picture of predicted project performance cannot easily be discerned. This can be overcome by combining information on schedule, budget, and the project deliverables. This is called earned value analysis.

Earned Value Analysis

The technique of *earned value analysis* (EVA) allows the actual performance of the project to be compared to the predicted performance. All the information required for this type of analysis should be available from standard project reporting against the schedule and budget (reporting is described later in this chapter). The schedule per work package (or, if preferred, discrete product) is plotted on one axis of a graph, and the budget per work package (or product) is plotted against the other axis. See Figure 1.10.

Common acronyms are used for the information required for the analysis:

- *BCWS*. Budgeted cost of work scheduled (how much money has been allocated to each work package or product in the schedule)
- *BCWP*. Budgeted cost of work performed (how much money was allocated to each work package or product that has been completed)
- *ACWP*. Actual cost of work performed (how much it actually cost for the work package to be performed or the product to be delivered)

The figure for budgeted cost of work performed (BCWP) is the earned value at the point in time that the analysis is being done. The chapter by Brandon looks at this technique in greater detail. At this point it is just worth noting the control information that can be determined from EVA. By manipulation of the data gathered from project performance reporting, project schedule and budget performance can be assessed. The Schedule Performance Index (SPI) and Cost Performance Index (CPI) are calculated as follows:

$$\text{behind schedule if } SPI < 1 \qquad SPI = \frac{BCWP}{BCWS} \qquad CPI = \frac{BCWP}{ACWP} \qquad \text{over budget if } CPI < 1$$

If, at the time of measurement, budgeted cost of work performed and the budgeted cost of work scheduled are the same (i.e., the SPI = 1), the project is exactly on schedule. In the

FIGURE 1.10. EARNED VALUE ANALYSIS PRESENTED GRAPHICALLY.

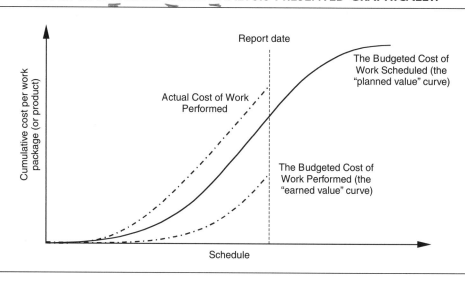

same manner, if the budgeted cost of work performed is the same as the actual cost of work performed (i.e., the CPI = 1) the project is exactly on budget.

If the BCWP is less than the BCWS, the SPI is less than 1; therefore, the project is behind schedule. Equally if the BCWP is less than the ACWP, the CPI is less than 1; there, the project is over budget.

EVA appears relatively straightforward to use, and predictions of future performance can be made using the data (by projecting final time and cost to complete using SPI and CPI values). However, care must be taken to moderate the results from this technique with other project data. There are also inherent dangers in believing that the information provided is a foolproof indicator of current and future progress. EVA does not report the subtleties of project control; it only provides an overview.

Key Performance Indicators

Key performance indicators (KPIs) are used to measure project progress toward achieving objectives, rather than the detail of progress of the work packages. They may be used to

- Measure project performance that is directly related to the change the project is delivering (which could be shareholder value, return on investment, market share, etc.)
- Measure project specific performance—that is, the performance of the project processes (e.g., effectiveness of project control mechanisms, degree of project cost reduction by using designated procurement practices, amount of change occurring in project, etc.).

KPIs must be determined at the beginning of the project and provide direct progress information toward project objectives. The information these measures of performance provide can help the project manager make decisions on trade-offs between the various (usually conflicting) control actions needed.

KPIs also need to be measurable (otherwise how will one know if they have been achieved?). Whilst this sounds obvious, it must be remembered that KPIs can only be useful if the information needed to determine the KPI during and at the end of the project is actually available. This implies that the project management information system must collect relevant data and generate the appropriate information outputs to provide the KPIs to the project's management team—upon which control action will be based.

If the KPIs to be used in a project have been determined by consultation between those needing the change to be delivered by the project (the project sponsor) and the project manager, it is possible to define success as meeting the KPIs at project completion.

Critical Success Factors

Critical success factors (CSFs) are sometimes used synonymously with KPIs. Literally, however, CSFs are the factors that are critical to success. Identification of the CSFs for a project will mean that the project manager and project team know where to concentrate their attention in order to achieve the project objectives. CSFs are therefore the factors that are critical to achieving success, *not* a measure of performance—which is what KPIs are.

A number of studies have been conducted into the factors found to be critical to project success. Many are generic across all projects, but each project will also have its own very specific factors. In their definitive research on success factors in projects Morris and Hough (1987) identified CSFs under the following general headings:

- Project definition
- Politics/social factors
- Schedule urgency
- Legal agreements
- Human factors
- Planning, design, and technology management
- Schedule duration
- Finance
- Project implementation

The development of CSFs for the project (with the involvement of the project manager, project team, project sponsor, and other senior stakeholders) is an important exercise in its own right, since all those associated with the project gain mutual understanding of what is critical to project success.

Performance Measurement and Control Action

The elements of the project plan have all now been described. On large and complex projects, these elements are often combined into a comprehensive project management control system. These systems also usually aggregate information from many projects, up to their respective program plan, if they are part of a program, and then up to the organization's business management system.

The project plan is constantly adjusted to reflect the reality of what is happening during the project, and so enable the effect of control actions to be predicted on the progress of the current and future work packages. The updated plan provides information to the following:

- *The project team.* Who can then plan their work packages to suit the revised plan
- *The project sponsor.* Who can assess the impact of the new plan on the delivery of the change required
- *Other project stakeholders.* Who may have other areas of work impacted by the changed project plans (including other projects being run—which is particularly important for program managers)

It is critically important, however, that the original plan is not lost—the project plan must be "baselined." This means that while the plan is updated and used to replan future work, it is still possible to compare what *should* have been completed (the baseline plan) with what *has* been completed (the updated plan). Knowledge of the variance between the two plans provides performance information and therefore helps

- Guide the development of the control actions required to bring the project back towards its original plan (if so desired)
- Improve the future control actions to make meeting the new plan more likely
- Gain knowledge to improve future planning (for replanning the same project and for plans for new projects)

The gathering of information to be used for project control is known as *project performance measurement.* The process provides an integrated view of the performance of the project—cost, schedule, technical issues, commercial, and business issues—so that control action can be taken where necessary to correct undesirable variances from the project plan. Equally important is the appropriate reporting of this information—at the right time, to the right people, and in the right format.

The measurement and reporting of progress must fundamentally begin at the work package level. It is here that the information for performance measurement originates. The work package managers must gather information on progress on the specific deliverables that they, or their team, are responsible for creating. The information must be presented in the same manner as the project plan presents it. In this way, variances from the plan are identified at the point where the variance occurs. The work package manager can then instigate control action to bring performance back in line with the project plan—normally a day-to-day management activity. (See Figure 1.3 for a reminder of how all the project control loops nest within a hierarchical control system.)

Performance information is reported to the project manager on a regular basis (weekly and monthly usually). Integrated reporting means that all the work package managers report the same measurements, together with the control actions they have taken to reduce negative variance from the plan. Hence, an aggregated project performance report can be compiled. (Sometimes project performance is reported on an exception basis; i.e., a report is generated only when there is a variance to the project plan requiring the attention of the project manager. Even in those organizations that use such reporting methods, a monthly reporting cycle is common.) From this report the project manager can determine which work packages are underperforming and whether the control action taken is likely to correct the situation. This reporting is the control feedback loop shown in the diagrams earlier in the chapter.

With the overview of project progress afforded by the integrated report, the project manager is in a position to assist the work package managers to improve performance in a manner that does not compromise other work package performance. This is important, since many work packages will be interrelated and may also share resources. The information can be used to replan work packages, and hence the project, and may also mean that the project manager can take action at a level above the work packages. Examples (amongst many possibilities) include the following: work packages could be rescheduled, the specifications of the work package deliverables may be changed, the acceptance criteria of the deliverables against the requirements may be adjusted, or the project scope may be changed.

This entire process is central to the notion of effective project management. The project manager is the single point of integrative responsibility. It is his or her principal function to integrate control action for the greater needs of the project, to ensure that objectives are met and that the desired change is created by the project. After the control action has been taken, the subsequent work package reports will provide evidence of whether variances from the plan have been successfully controlled—and so the process continues.

Organizations often require visibility of project performance at a program or portfolio level. This enables better management of organizational budgets and control of the changes being created by multiple projects and programs. "Rolled-up" performance measurement information, often supported by a program support office (see the chapter by Young and Powell), enables summary reporting to be available to appropriate levels of management in the organization.

Summary

This chapter has outlined the processes that constitute project control, and in so doing introduced many other project processes upon which effective control depends.

Fundamentally, the project must have a

- Set of objectives directly related to the need for the change the project is set up to deliver
- Plan against which the project can be controlled—and so deliver those objectives
- Process to measure performance against the plan—the feedback loop
- Process to control changes to the scope
- Project manager who is truly the single point of integrated control action, with responsibility for delivering the change required of the project

The control process goes on throughout the project life cycle because the internal and external environment of the project is continuously changing. For example:

- Performance of resources is often other than predicted (better or worse).
- New information is generated that may indicate the original plan was not feasible to begin with.
- The objectives for the project may change because the change required to be brought about (by running the project) is itself changed.

In combination, these processes can be seen as forming the "iron rules" for the project. Project control is about

- Good planning of scope, schedule, and budget
- Setting up appropriate metrics to monitor performance

- Reporting the performance against those metrics
- Replanning and instigating corrective action to reduce variance from the baseline plan

References

Association for Project Management. 2000. *Body of Knowledge 4th Edition*. High Wycombe, UK.

Allinson, K. 1997. *Getting there by design*. Oxford, UK: Architectural Press.

Barnes, M. 1988. Construction project management. *International Journal of Project Management* 6 (2, May): 69–79

Bertalanffy, L., von. 1969. *General systems theory: Essays on its foundation and development*. New York: Braziller.

Checkland, P. 1981. *Systems thinking, systems practice*. Chichester, UK: Wiley.

Cleland, D. I., and W. R. King. 1983. *Systems analysis and project management*. International ed. Singapore: McGraw-Hill.

Dixit, A. K., and B. J. Nalebuff. 1991. *Thinking strategically: The competitive edge in business, politics, and everyday life*. New York: W. W. Norton & Co.

Dixon, M. 2000. *Project management body of knowledge*. 4th ed. High Wycombe, UK: The Association for Project Management.

Eisner, H. 1997. *Essentials of project and systems engineering management*. New York: Wiley.

Jackson, T. 1993. *Organisational behaviour in international management*. Oxford, UK: Butterworth-Heinemann.

Katz, D., and R. L. Kahn, 1969. Common characteristics of open systems. In *Systems thinking*, ed. F. E. Emery. 86–104. Harmondsworth, UK: Penguin Books.

Morris, P. W. G. 1994. *The management of projects*. London: Thomas Telford.

Morris, P. W. G., and G. H. Hough. 1987. *The anatomy of major projects*. Chichester, UK: Wiley.

Pinto, J. K., and D. P. Slevin, 1988. Critical success factors across the project life cycle. *Project Management Journal*. 19(3):67–75.

Project Management Institute. 2000. *A guide to the project management body of knowledge*. Newtown Square, PA: Project Management Institute.

Quinn, J. B. 1978. Strategic change: Logical incrementalism. *Sloan Management Review*. (Fall) 7–22.

Reinertsen, D. G. 1987. Managing the design factory. New York: Free Press.

Turner, J. R., and D. Remenyi. 1995. Estimating costs and revenues. In *The commercial project manager* by J. R. Turner. 31–52. London: McGraw-Hill.

Suggested Further Reading

Archibald, R. D. 2003. *Managing high technology programs and projects*. New York: Wiley.

Hartman, F. 2000. *Don't park your brain outside*. Newtown Square, PA: Project Management Institute.

Murray-Webster, R., and M. Thiry. 2000. Managing programmes of projects. In *The Gower handbook of project management*. 3rd ed, *ed.* R. Turner, S. Simister, and D. Lock. Aldershot, UK: Gower.

Smith, N. J. 2002. *Engineering project management*. 2nd ed. Oxford, UK: Blackwell Science.

CHAPTER TWO

TIME AND COST

Asbjørn Rolstadås

Time and cost are two important planning and control variables in a project. They are interdependent. For example, an acceleration of a schedule may lead to reduced productivity on the work carried out and thus to increased costs or it may require resources that only are available at extra costs (such as overtime). A delay or a prolongation of a project may also involve extra costs to carry the project management and administration for the extra time.

In this chapter scheduling will first be discussed without taking resource constraints into account. The traditional network scheduling techniques such as CPM (critical path method) and PERT (program evaluation and review technique) will be explained by use of examples. Then techniques for handling resource constraints will be briefly discussed.

Scheduling Representation

All projects will have to comply with a deadline for their finish. For a project owner, the deadline may be set to comply with the needs of the results of the project. Quite often the deadline is important in the project's overall profitability analysis. An extension to the deadline may in the worst case turn a profitable project into a nonprofitable one. Deadlines may also be set by external conditions such as weather. As an example, offshore oil platforms for the North Sea are fabricated on shore and can only be towed to field during the summer because of the risk of bad weather in the winter time. For a contractor the deadline is usually a part of the contractual conditions and may involve penalties if not met.

This section of the chapter discusses how it is decided when each single activity of the project should be executed in order to meet a predetermined deadline or when a project can be expected to be finished given that the duration of each activity is known. This is referred to as scheduling and represents the process of determining when an activity should start and when it should finish.

There is a rich literature available on the scheduling of projects. Almost all discuss bar charts and networks. The difference is how deep into network techniques they go and whether or not they consider resource constraints. Some general references are given at the end of the chapter.

In scheduling a distinction is made between the following:

- Activities
- Events

An *activity* is defined as a number of job assignments that requires resources to be accomplished. An *event* is a point in time where an activity starts or ends. Hence, there is a duality between activities and events. For any activity, there are two associated events (start and stop), and for any two adjacent events, there is one, and only one, activity connecting them.

Milestones are a special type of events. A *milestone* represents an event against which achievement is measured. Milestones are used to control progress at a high level. They are selected as major completion points or decision points. A milestone that represents a decision point is also referred to as a *decision gate* or a *tollgate*. The idea is that before the project is allowed to proceed, the gate must be passed. The gatekeeper checks that necessary documentation exists and that the required preceding tasks are satisfactory completed.

The most widespread scheduling tool is the Gantt chart (or bar chart). Henry Gantt, who tried to schedule logistics for the war front in World War I, invented it. The Gantt chart is very simple. It is a rectangular diagram with time on the horizontal axis and activities on the vertical axis. In the diagram, bars show the timing and duration of activities. An example is given in Figure 2.1.

The Gantt chart is unsurpassed in its efficient way of communicating a schedule. Any reader can understand and read the diagram without any prior training or knowledge. However, there are a few drawbacks with the diagram:

- It does not include information on resource level for the activity. It just states when the activity starts and when it finishes.
- It does not include any information on precedence relationships.

For example, with reference to Figure 2.1, if the activity "deck mating" should start earlier, the diagram does not show that "GBS fabrication" and "deck fabrication" must also be moved forward. However, there is a revised version of the original Gantt chart, referred to as a *linked Gantt chart*, that includes this information.

The Gantt chart can also be used for monitoring schedule progress, as shown in Figure 2.1. The open bars indicate unfinished activities. As the activity progresses, the bar is filled. A vertical (dashed) line indicates the present day. A quick look at the diagram reveals crucial progress information. In the figure, it is easily detected that "GBS fabrication" is seriously behind schedule.

FIGURE 2.1. EXAMPLE OF A GANTT CHART.

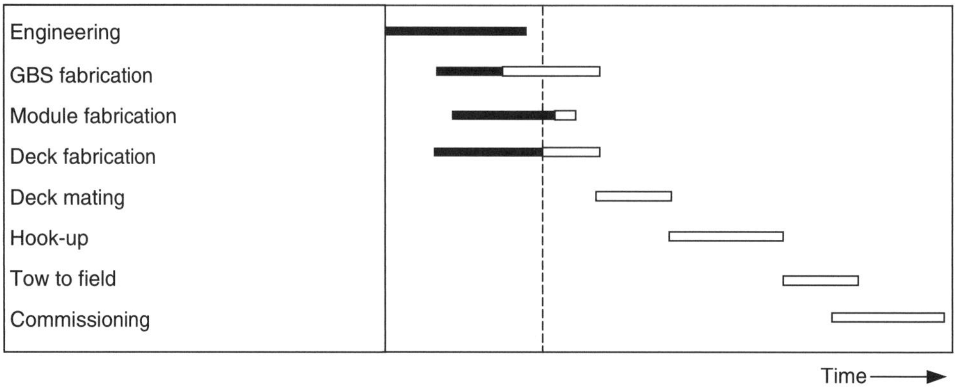

Engineering
GBS fabrication
Module fabrication
Deck fabrication
Deck mating
Hook-up
Tow to field
Commissioning

Time →

Network Representation

Even though a linked Gantt chart can include precedence relationships between activities, it is an unsuitable tool if these relationships grow beyond a certain complexity. In such situations, the scheduling should be carried out by a network technique. However, the presentation of the schedule to the project participants may still be done in a Gantt chart format. Most project control packages available offer this flexibility.

There are two types of network representations:

- *Activity on arrow* (AOA)
- *Activity on node* (AON)

Figure 2.2 shows an example of an AOA representation. The arrows of the graph represent the activities. The arrows also define the precedence relationships, for example, "D and E must be finished before H may start." The nodes in the AOA network represent events.

In AON representation the very same project is shown in Figure 2.3. The nodes of the graph represent the activities. The arrows give only precedence relationships. Events are not shown in AON networks.

In order to design a network, basically three types of information are needed:

- A list of all the activities
- A list of all precedence relationships
- An estimated duration of each activity

The list of activities and the precedence relationships define the structure of the network. Table 2.1 gives an example of these data for the network shown in Figures 2.2 and 2.3. The precedence relationship is given by defining the immediate preceding activities. For

FIGURE 2.2. EXAMPLE OF AN AOA NETWORK.

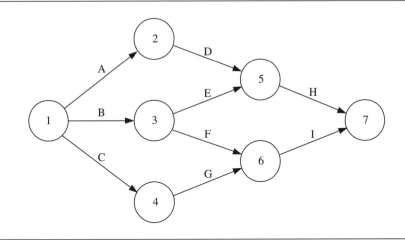

FIGURE 2.3. EXAMPLE OF AN AON NETWORK.

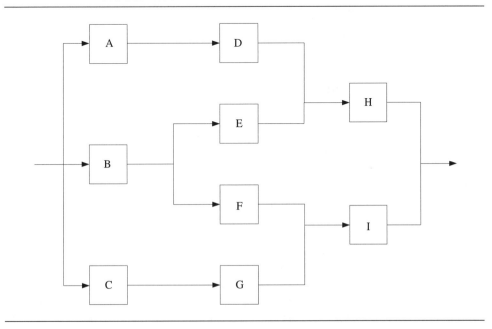

TABLE 2.1. DATA DEFINING STRUCTURE OF NETWORK IN FIGURES 2.2 AND 2.3.

Activity	A	B	C	D	E	F	G	H	I
Preceding activity	—	—	—	A	B	B	C	D, E	F, G

example, for activity H, the preceding activities are D and E. This means that D and E must be finished before H can start.

Constructing the Network

During project scoping, a work breakdown structure (WBS) is usually developed. During scheduling, each work package normally represents one activity in the network. The schedule serves as a baseline for time control and is usually referred to as a *control schedule*. The one that is approved for execution start is referred to as *the master control schedule* (MCS).

When the project execution starts, invariably there will soon be a need for changing and updating the schedule. Since the control schedule is a reference document for monitoring and control, it should at any time be as realistic as possible. This requires that it be kept updated. Usually it is updated on a periodic basis, depending on variability of the schedule and the need for up-to-date control information. The reference schedule at any time is referred to as the *current control schedule* (CCS). Sometimes a number is included to indicate the update number; for example, CCS (3) is the third update of the schedule.

Since the schedule is directly connected to the WBS, aggregate schedules may be obtained by aggregation to a higher WBS level than the work package. For example if the WBS levels are

$$\text{Project} \rightarrow \text{Contract package} \rightarrow \text{Work package}$$

a master schedule with contract packages as activities may be derived from the CCS by aggregation.

The purpose of a network scheduling is twofold:

- To determine when the project will finish (its duration)
- To determine which activities directly influence the project duration

Network scheduling was developed during the 1950s. Two techniques were developed in parallel:

- Program evaluation and review technique (PERT)
- Critical path method (CPM)

The main difference between the two techniques is how the duration of the activity is estimated. In PERT networks the duration is a stochastic (uncertain) variable following a statistical distribution. In CPM, the duration is deterministic (certain). PERT is therefore able to handle uncertainty and may give answers to questions like "What is the probability of finishing the project by March 12" or "What date should be given for milestone X if there should be more than 90 percent probability of meeting it." CPM does not handle uncertainty. It assumes that the duration is a fixed number.

CPM Networks

CPM calculates a network in several steps:

- For each event:
  - *Earliest possible time.* The earliest possible time the event can occur
  - *Latest possible time.* The latest possible time the event can occur
- For each activity:
  - *Early start (ES).* The earliest possible start time for the activity
  - *Early finish (EF).* The earliest possible finish time for the activity
  - *Late start (LS).* The latest possible start time for the activity
  - *Late finish (LF).* The latest possible finish time for the activity
  - *Float (FL).* The amount of time an activity may be delayed compared to early start without jeopardizing the project deadline

The calculation of events is done in a forward (for early times) and a backward (for late times) "pass." In the forward pass all possible activities leading into an event are examined and an event time is calculated along each of them. The maximum value is chosen. Early event time for event i can be computed as:

$$e_i = \max_j (e_j + t_k)$$

$$j \in PE_i, k \in PA_i$$

where:

e_i is early event time for event i.
t_k is duration of activity k.
PE_i is set of directly preceding events to event i.
PA_i is set of activities having event i as finish event.

For the backward pass, the late event times are determined in a similar way:

$$l_i = \min_j (l_j - t_k)$$

$$j \in SE_i, k \in SA_i$$

where:

> l_{li} is late event time for event i.
> SE_i is set of all directly succeeding events to event i.
> SA_I is set of all activities having event i as its start event.

Figure 2.4 shows the results of the calculations for the network in Figure 2.2. The activity durations are taken from Table 2.2 and are shown on the network in parentheses after the activity name. How durations can be calculated is shown in the *Resource Constraint* section later in the chapter.

For example, for the calculation of early event time of event 6, it is realized that there are two preceding events (3 and 4) with early event time of 7 and 12, respectively. Following activity F from event 3 to 6 would give an early event time of $7 + 13 = 20$ for event 6. Following G from event 4 to 6 would likewise give $12 + 16 = 28$. The larger number of 20 and 28 (28) is then selected as the early event time for event 6, since the event can occur only when both activities F and G are finished.

For the activities the following calculations are done:

FIGURE 2.4. CPM CALCULATIONS FOR AN AOA NETWORK.

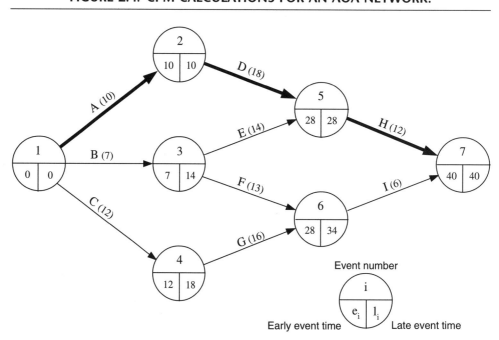

TABLE 2.2. ACTIVITY DURATIONS FOR NETWORK IN FIGURE 2.4.

Activity	A	B	C	D	E	F	G	H	I
Duration (weeks)	10	7	12	18	14	13	16	12	6

$$ES_i = e_e$$

$$EF_i = ES_i + t_i$$

$$LF_i = l_l$$

$$LS_i = LF_i - t_i$$

$$FL_i = LF_i - EF_i$$

where:

e_e is earliest possible time for start of event for activity i.
l_l is latest possible time for finish of event for activity i.

The calculations for the network in Figure 2.4 are shown in Table 2.3.

The term "float" may need some further explanation. With reference to Figure 2.4, there is a time window for execution of activity G limited by the early time of event 4 (12), and the late time of event 6 (34). The time window spans a total of $34 - 12 = 22$ days. The duration of G is 16 days, which leaves a surplus of $22 - 16 = 6$ days. These 6 days represent a freedom in scheduling of activity G. The start of G may be delayed by up to

TABLE 2.3. CALCULATION OF ACTIVITY TIMES FOR NETWORK IN FIGURE 2.4.

Activity	Duration	ES	EF	LS	LF	FL
A	10	0	10	0	10	0
B	7	0	7	7	14	7
C	12	0	12	6	18	6
D	18	10	28	10	28	0
E	14	7	21	14	28	7
F	13	7	20	21	34	14
G	16	12	28	18	34	6
H	12	28	40	28	40	0
I	6	27	33	34	40	6

six days without affecting the finish time of the project. Activities with no such freedom (float) are denoted critical activities. A chain of critical activities from start to finish in the network is called the *critical path*. In Figure 2.4 the critical path is A-D-H.

The concept of critical activities is quite interesting. It draws the attention of the project manager to the activities that need the closest monitoring. Any delay of a critical activity leads to an equivalent delay of the total project.

Usually activities are scheduled for start as early as possible (ES). If this is the case, the entire float can be taken up by a delay without affecting the finish time of the project. However, it must be kept in mind that the float is valid for a chain and not a single activity. Again with reference to Figure 2.4, the chain of activities C and G together have a float of 6 days. That means that the sum of delays for C and G may run up to 6 days without affecting the project finish time.

Analysis of float is a particularly neat tool for calculating consequences of schedule variance. An example (with reference to Figure 2.4 and Table 2.3) may help to understand this. Assume the following field report with respect to schedule:

- B will be delayed by 4 days.
- D will be delayed by 1 day.
- E will be delayed by 5 days.
- G will be delayed by 3 days.

It is recognized that D is critical. Hence, a delay of at least 1 day to the overall project is unavoidable. Activity G has a float of 6 days. Since no other activity on that chain has a delay, the float will accommodate the 3-day delay of G, and this delay will therefore not influence the project finish date. Further, B and E are both on the same chain. The float along this chain is 7 days, and the total delay is $4 + 5 = 9$ days. This means a two-day delay of the project. In conclusion the project will be delayed by 2 days and B-E-H will be the new critical path. A-D will have a float of 1, and C-G a float of 4.

The CPM calculation of an AON network is similar to that of an AOA network, but the calculation of events is omitted. Figure 2.5 shows these calculations for the same network (first shown in Figure 2.3).

The calculations are done in two passes as for event calculation of an AOA network. In the forward pass, all early start and finish times are determined:

$$ES_i = \max_j(ES_j + t_j) \, j \in PA_i$$

$$EF_i = ES_i + t_i$$

where PA_i now is defined as the set of all directly preceding activities of activity i.

For the backward pass the following calculations are done:

FIGURE 2.5. CPM CALCULATIONS FOR AN AON NETWORK.

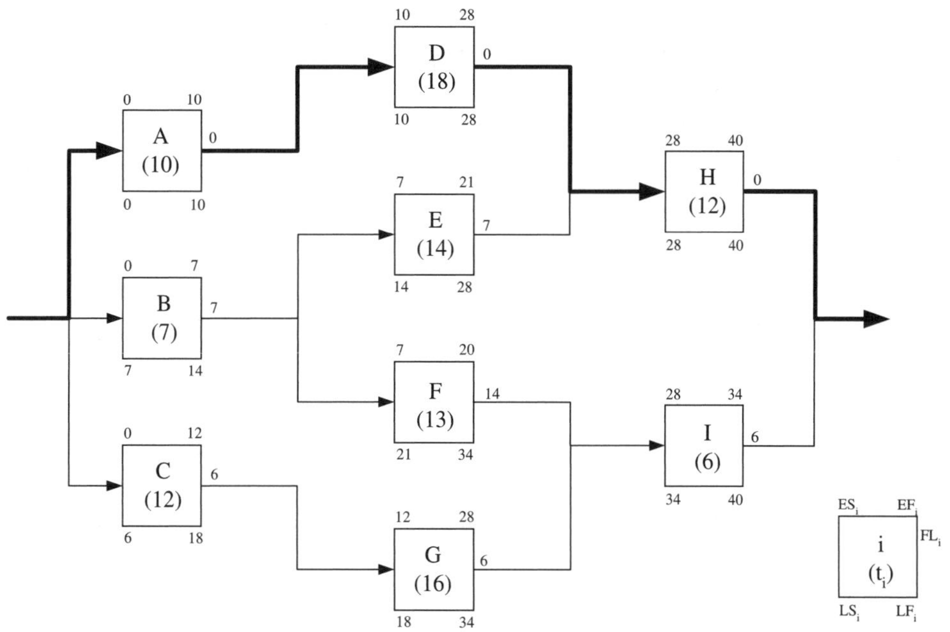

$$LF_i = \min_j(LF_j - t_j)\, j \in SA_i$$
$$LS_i = LF_i - t_i$$

where again SA_i now denotes the set of all directly succeeding activities to activity i.
    Float is calculated as before.

Precedence Networks

In traditional network theory only one type of precedence relationship applies: "Activity B may start as soon as activity A is finished." This is called a "finish to start" precedence relationship. The finish of one activity is related to the start of some other activity.
    Naturally, there may in principle be four different types of precedence relationships, as indicated in Figure 2.6:

- Finish to start (FS)
- Start to start (SS)
- Finish to finish (FF)
- Start to finish (SF)

FIGURE 2.6. PRECEDENCE RELATIONSHIPS BETWEEN TWO ACTIVITIES.

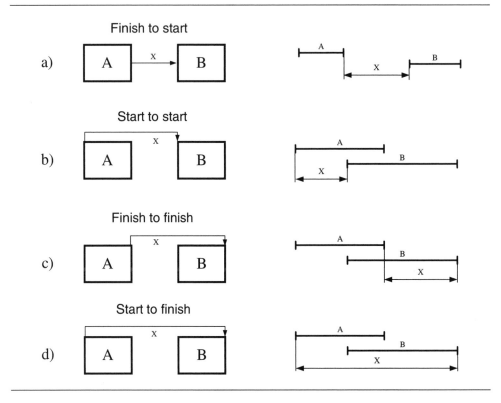

Further, a time delay of x units may be added. An FS$_x$ relationship between A and B means that B can start x time units after the finish of A. For an SS relationship, the start of A determines the start of B. A and B may run in parallel with a time phase of x time units. The same is true for an FF relationship with the exception that it is the finish of the two activities that are time phased rather than the start.

The SF relationship, on the other hand, defines a total time window of x time units for the activities. As an illustration, assume that activity A is "engineering design of a petro-chemical process" and that activity B is "procurement of equipment items." Obviously, in practice one may allow these activities to overlap. However, they cannot overlap fully. To procure equipment, a certain amount of engineering design must be completed, say, 30 percent representing a duration of 8 weeks. This means that there is a start-to-start rela-tionship between A and B with an 8-week (30 percent) time lag (SF$_8$).

Also, in this example, the procurement needs some time after the finish of engineering design to do the last purchasing, say, about 3 weeks. This means that there is also an FF$_3$ relationship between activities A and B.

Any of these activities may be combined with each other. This provides a good flexibility to define a network with different overlaps of activities. Many say this is more realistic.

However, there is also a danger to start adapting the overlap conditions to meet a desired project due date. In this way the project manager may be fooled into overlap conditions that are unrealistic and that will show during project execution.

Precedence networks are calculated very much in the same way as traditional networks. It is beyond the scope of this chapter to go into this, and readers are encouraged to reference other literature on this topic.

PERT Networks

PERT allows activity duration with uncertainty. It assumes a statistical distribution for the duration of each single activity.

The most commonly applied distribution is a β-distribution. The β-distribution is finite between a and b and has a mode (most probable value) of m. a and b are interpreted as the most optimistic and pessimistic estimate of the duration. m is the most likely duration. By selection of the parameters a, b, and m, almost any skewed distribution may be constructed. Usually the distribution is skewed to the right, indicating that a delay is more likely than a finish ahead of scheduled due date.

The PERT method assumes that all activity durations are stochastically independent. This means that there is no underlying connection leading to simultaneous duration variations of two or more activities. If activity X is delayed, this does not lead to a similar delay of activity Y. This is, of course, a questionable assumption. Quite often there are activities that are connected (covariance). For example, a delay of an activity may be caused by a shortage of labor. In this case it is reasonable to believe that this is the case also for other activities performed in the same region.

PERT calculates expected durations for all activities and then does an ordinary CPM calculation of the network using these expected values as durations. For the β-distribution, the expected value and variance may be calculated as:

$$E(t_i) = \frac{1}{6}(a_i + 4m_i + b_i)$$

$$Var(t_i) = \frac{1}{36}(b_i - a_i)^2$$

where:

$E(t_i)$ is expected value of activity i duration.
$Var(t_i)$ is variance of activity i duration.
a_I is optimistic estimate of activity i duration.
b_I is pessimistic estimate of activity i duration.
m_I is realistic estimate of activity i duration.

The variance is a measure of the uncertainty of the duration. The larger variance, the larger is the uncertainty.

If p is a path (a chain of activities) in the network, the total duration along that chain is

$$T_p = \sum_{i \in p} t_i$$

This is, of course, trivial, since it just says that the duration of the path is the sum of the durations of the activities along the path. If the path is the critical path, π, the duration, T_π, equals the project duration.

With the assumptions made, the expected value and variance for T_p can be computed as follows:

$$E(T_p) = \sum_{i \in p} E(t_i)$$

$$Var(T_i) = \sum_{i \in p} Var(t_i)$$

Even though each t_i follows a β-distribution, T_p, may be approximated with a normal distribution. This is convenient, since the normal distribution is readily available in a table format.

An example may clarify the use of PERT networks. Assume the same network as earlier (see Figure 2.5). Table 2.4 gives the three estimates (a, b, m) for all the activities. In practice, these estimates have to be provided, either based on experience data or an assessment of the uncertainty of the activity. In the table, expected values are calculated for all the activities and the variance is calculated for the critical activities. Note that the most likely estimate (m) is the same as the duration used for the CPM calculation.

The network is now calculated using the expected values as shown in Figure 2.7. The new total duration is 44 days instead of the 40 days that the CPM calculation gave. The 4-day difference is a measure of the risk associated with each activity's duration. Since for most activities a delay is more possible than a finish ahead of schedule, the expected value

TABLE 2.4. DURATION ESTIMATES FOR A PERT NETWORK.

Activity	m	a	b	E(t)	Var(t)
A	10	9	17	11	1.78
B	7	5	9	7	0.44
C	12	10	20	13	2.78
D	18	16	32	20	7.11
E	14	13	21	15	1.78
F	13	10	16	13	1.00
G	16	15	23	17	1.78
H	12	11	19	13	1.78
I	6	5	7	6	0.11

FIGURE 2.7. EXAMPLE OF A PERT NETWORK CALCULATION.

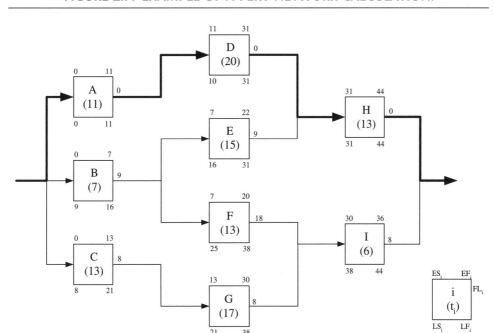

is somewhat higher than the most likely value. For example, for activity A the most likely value is 10 days, but the expected value is 11 days. The one extra day compensates for the risk of overrun.

The critical path is as before A-D-H. The duration of the project is

$$T_\pi = t_A + t_D + t_H$$

and expected value and variance is

$$E(T_\pi) = E(t_A) + E(t_D) + E(t_H) = 11 + 20 + 13 = 44$$

$$Var(T_\pi) = Var(t_A) + Var(t_D) + Var(t_H) = 1.78 + 7.11 + 1.78 = 10.67$$

The variance as such does not provide much practically useful information about uncertainty. It has to be transferred to a decision parameter that the project manager can understand and use. One way to do this is to calculate the probability of finishing the project on or before a given date or the probability of meeting a milestone that has been set. For example, the project manager would like to know what the probability is of reaching the 40-days deadline calculated by the CPM method. The calculations are as follows:

$$\Pr(T_\pi \leq 40) = \Phi\left[\frac{40 - E(T_\pi)}{\sqrt{Var(T_\pi)}}\right] = \Phi\left(\frac{40 - 44}{\sqrt{10.67}}\right) = \Phi(-1.23) = 0.109$$

where Φ is the normal distribution (0.1).

As can be seen, the chance of meeting this deadline is not very large, only 10.9 percent. The project manager can rephrase the question and ask what deadline he or she should give if he or she wants a 90 percent probability of finishing on or before the deadline. If this deadline is D, the calculations are as follows:

$$\Pr(T_\pi \leq D) = \Phi\left(\frac{D - E(T_\pi)}{\sqrt{Var(T_\pi)}}\right) = \Phi\left(\frac{D - 44}{\sqrt{10.67}}\right) = 0.90 = \Phi(1.28)$$

$$\frac{D - 44}{\sqrt{10.67}} = 1.28 \Rightarrow D = 48.2$$

So in order to have this at least 90 percent guarantee against a delay, the deadline quoted should be 49 days.

In the calculations $\Phi(x)$ is taken from a table of the normal distribution. This may be found in any statistical tables book.

Earlier, the assumption that all activity durations are stochastically independent was discussed. It was said that this assumption might be unrealistic. There is another assumption with the PERT method that is also quite unrealistic. Assume that there is a path other than the critical path that is nearly critical. For example, it may have a float of 1 of 2 days. Since durations may vary stochastically, there may be a situation where one of the activities on the critical path becomes very short at the same time as one activity on the other path becomes very long. This may cause the critical path to change from the original one to one that was nearly critical. PERT does not handle such situations, since all calculations are done under the assumption that the critical path does not change.

This may be one reason why the PERT method is not frequently applied in practice. Another reason is, of course, the problem of making the three estimates for activity duration (a, b, and m). In any case, a computer will do calculations. Then it may be just as easy to do a Monte Carlo simulation.

Monte Carlo Simulation

A Monte Carlo simulation will use the same input data as PERT—that is, statistical distribution for the duration of each activity. Then it will draw a set of durations for all activities from its statistical distribution. Then the network is calculated as a usual CPM network with these durations. The procedure is repeated a large number of times (for example, 1,000), and in this way an empirical statistical distribution is achieved. This may again be used to answer questions of what is the probability of meeting a milestone or a due date.

If a Monte Carlo simulation is done, there is a very interesting by-product. By counting the number of times an activity is critical, the probability of an activity being critical can be estimated. This is referred to as the *critical index* of the activity, and it may be a powerful decision parameter for the project manager. CPM only says whether an activity is critical or not. For example, A may be critical, but B is not. If A has a critical index of 51 percent and B a critical index of 49 percent, it is seen that both A and B may be critical with almost the same probability (51 percent versus 49 percent).

If a project scheduling needs to take uncertainty into account, a Monte Carlo simulation is probably the best approach. There are commercial products available for doing such analysis.

Resource Constraints

In the preceding discussions, resource constraints have not been taken into account. In practice, resource constraints, of course, apply in a number of ways:

- To determine the duration of an activity
- As a trade-off between time and cost
- As a limited resource during scheduling

Each of these three cases is briefly discussed in the succeeding paragraphs.

When the project is being scoped, an estimate of work volume is done, for example, as a number of person-hours for an activity (work package). If the activity is labor-intensive, it is common to apply a buildup and a rundown period to and from peak manning of the activity. This may be referred to as *mobilization* and *demobilization*. In this case the resource profile looks like a trapezoid as shown in Figure 2.8. It is assumed that the work scope is known and that the peak manning level and the transfer mobilization and demobilization are given. Realizing that the area of the trapezoid equals the work scope, the activity duration, t, can be calculated as:

FIGURE 2.8 THE TRAPEZOIDAL METHOD.

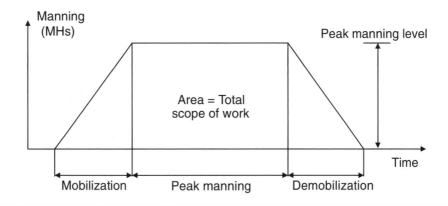

$$t = t_m + t_p + t_d$$

where:

t is activity duration.
t_m is mobilization time.
t_p is time with peak manning level.
t_d is demobilization time.

Then the following calculations can be done:

$$S = c \cdot P_{\max} \frac{t_m + t_d + 2 \cdot t_p}{2}$$

where:

S is scope of work (person-hours).
c is net person-hours per person per week.
P_{max} is peak manning level.

By solving this equation with respect to t_p and inserting in the expression for t, the trapezoidal formula is derived:

$$t = \frac{S}{c \cdot P_{\max}} + \frac{1}{2}(t_m + t_d)$$

This formula gives the duration as a fixed number. In practice, the duration may always be influenced. For example by the use of overtime or extra manpower, the activity may be accelerated. This would then involve a higher cost. This again allows for a trade-off between time and cost, as shown in Figure 2.9.

The curve shows the normal duration and its associated cost. It further indicates that the activity may be accelerated (duration shortened) until a limit referred to as "crash duration." The corresponding costs are called "crash costs."

The time/cost relationship allows formulating the selection of duration as an optimization problem. There will therefore be an optimal duration. The mathematical approach to this is called *operations research* and has found wide application among researchers but is hardly used in practical project planning and control. There may be various reasons for this; an important one may be the complexity of the problem and that the model the optimization is based on most often is quite unrealistic.

In a practical situation one would start with a schedule based on normal durations. Then one would use a heuristic strategy to reduce this as much as is required. The most common strategy is to calculate the cost gradient for each activity:

FIGURE 2.9. TIME/COST TRADE-OFF.

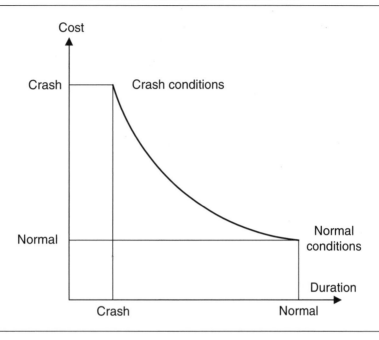

$$c' = \frac{c_D - c_d}{d - D}$$

where:

c' is cost gradient.
c_d is normal cost.
c_D is crash cost.
d is normal duration.
D is crash duration.

Then one would look at the critical path activities and reduce the one with the lowest cost gradient as much as possible. If this is not sufficient, one carries on with other activities in increasing sequence of the cost gradient value. Of course, it would make little sense to crash noncritical activities, because that would only buy extra float in the network.

If the resources are limited, the scheduling problem is referred to as "scheduling under resource constraints." In this connection there are really two variables that are considered:

• The project duration
• The resource level

Trying to fix both may create a problem with no solution. For example, if two welding activities running in parallel requires 2 and 3 welders, respectively, and the total number of available welders is 4, the only two solutions are either to obtain another welder (increase resource level) or to prolong the duration of one of the activities. Depending on the amount of float on that activity, the prolongation may involve longer project duration. Therefore, a standard approach is first to calculate the schedule without resource considerations and then adjust the schedule under resource limitations, keeping either the resource level or the project duration fixed and allowing the other to vary.

Scheduling with resource constraints is difficult and is a combinatorial problem of a magnitude that makes any exhaustive search (check all possible alternatives) impossible, even with the largest computers. However, a number of heuristic algorithms have been developed. It is beyond the scope of this chapter to discuss such algorithms, and you are referred to specialized books on scheduling or operations research.

Cost Management

In this section planning and control of costs are discussed. The cost estimate and its contingencies are introduced, and some basic rough estimating techniques are briefly explained. Then estimate updating and cost control is discussed. For a more complete picture of cost control including performance measurement by the earned value method, you are to the chapters by Harpum and by Brandon.

The Cost Estimate

The cost estimate is a forecast of the final project cost broken down on work packages and specified according to a code of account. This means that every estimate carries a certain amount of uncertainty. An estimate shall serve as a baseline for cost control. A fair estimate will have equal probability of overrun and underrun.

Any estimate will carry a contingency. There are two types on contingencies:

- *Contingency allowance.* Money to cover, normal, expected variances
- *Contingency reserve.* Money to cover unexpected variances

The *contingency allowance* should cover costs that are likely to occur but cannot at the present time be identified. The contingency allowance is meant to be spent in a normal project. It is not intended to cover unexpected events. A contingency to cover unexpected events is usually referred to as a *contingency reserve* and should cover costs that are unlikely to occur. It is intended not to be spent.

If each cost item has an associated uncertainty, the contingency allowance may be calculated. The uncertainty of the cost item can be estimated in the same way as for activity duration—that is, by giving optimistic, realistic, and pessimistic estimates and assuming a β-distribution.

Since the distributions normally will be skewed, the expected value will be higher than the most likely value. The sum of the most likely value for each cost item is referred to as a *base estimate*. Figure 2.10 shows the estimate value as a function of the probability. The amount that has to be added to the estimate to bring it up to a 50 percent probability may be interpreted as the contingency allowance. The contingency reserve is the additional amount needed to bring it up to, for example, 70 percent.

Usually, there are several estimates in a project. A distinction is made between estimates made prior to the start of project execution and estimates made during project execution.

Estimates made prior to the project start of execution should be classified according to at which project phase it has been made. This will allow an identification of estimate with respect to the following:

- The phase at which the estimate was made
- The documentation it is based upon
- The uncertainty the estimate carries

After project execution has started, the estimate will be updated to include approved variances at regular intervals. As with the schedule, the first estimate is denoted *master control estimate* (MCE) and the latest update is called *current control estimate* (CCE).

FIGURE 2.10. CALCULATION OF ESTIMATE CONTINGENCIES.

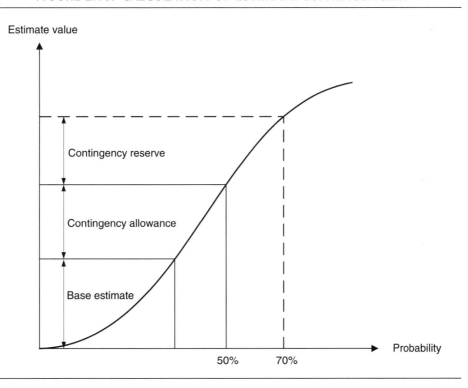

Estimating Techniques

Cost estimates are made at different stages in a project. At an early stage an estimate has to be provided before all technical details or design are developed. At a later stage more accurate estimates are made based on detailed design or bill of quantities. Developing the estimates at different stages will require different estimating techniques.

Basically there are two sets of estimating techniques:

- Synthetic
- Analytic

The synthetic techniques are used to obtain rough estimates. A project or work breakdown structure does not exist. The analytic techniques are based on a known WBS. Detailed estimates are made for each work package in accordance with a code of accounts. The analytic techniques are used for detailed estimates and are usually brought down to an uncertainty level of ± 10 percent. The synthetic techniques may be brought down to approximately $\pm 20\%$ uncertainty. They fall in two categories:

- Relational estimation
- Factor estimation

Relational estimation takes the cost of a finished similar project and corrects this for the following:

- Capacity of the facilities
- Time
- Location where the work is done

Factor estimation assumes that the percentage distribution of all cost items is known. Then one category is estimated in detail—usually the equipment—and the others are determined using the percentage distribution.

Both factor estimation and relation estimation are frequently used in practice. Their main application is selecting projects at a very early stage. Then the selected project will need an analytical estimate to be used as reference.

Cost Control

During project execution, costs need to be monitored and controlled (in addition to scope of work and time). The purpose of a cost control is to

- make project management aware of possible cost overrun at an early stage;
- inform all project team members of the current control estimate and the budget frame the work against; and
- establish cost consciousness amongst all project team members.

A full cost control in a project comprises the six steps indicated in Table 2.5. The first three are the traditional ones that are used in monitoring cost. The last three serve the purpose of really managing costs by analyzing the development and implementing corrective actions.

To implement the first three steps of the approach outlined in Table 2.5, a form like the one presented in Table 2.6 may be applied. Contingency allowance is used to cover variations, while contingency reserve is used to cover changes in scope of work.

The master control estimate (MCE) is updated with approved scope changes and variations to the current control estimate (CCE). To monitor the costs, both accrued expenditures and future commitments are registered. In addition, an independent forecast to complete is made. The sum of expenditures, commitments, and forecast to complete represents a revised forecast. This will be further analyzed through the Steps 4 to 6 in the approach outlined in Table 2.5. If the changes involved are approved, they will be included next period in the next update of the CCE.

Summary

In this chapter time and cost has been discussed. Together with the scope of work they make up the main control variables of a project. They are represented as two plan documents:

- Project schedule
- Project cost estimate

Together with the scope definition these two documents make up the project control baseline.

The purpose of these two documents is to clarify when the activities of a project should be executed and what the budget limit is for each of them. In addition, they serve as the reference for schedule progress and cost control of the project.

TABLE 2.5. THE SIX STEPS OF COST CONTROL.

Type of Activity	Step	Activity	Purpose
Cost monitoring	1	Know control estimate.	Know what to do.
	2	Keep account of commitments.	Know what has been done.
	3	Estimate costs to complete.	Know what remains.
Cost management	4	Analyze cost deviations.	Identify problem and its cause.
	5	Take corrective actions.	Minimize cost overrun.
	6	Develop revised forecast.	Estimate effect of actions.

TABLE 2.6. EXAMPLE OF A COST CONTROL FORM.

Cost Item	MCE	Scope Changes	Variations	CCE	Expenditures	Commitments	Forecast to Compl.	Revised Forecast
Contract A	2500	300	100	2900	400	900	1800	3100
Contract B	4000		500	4500	2000	300	2200	4500
Contract C	3500	100		3600	800	1500	1200	3500
Cont. allow.	2000	−400		1600				1500
Total estimate	12000		600	12600				12600
Cont. reserve	1500		−600	900				900
Total budget	13500			13500				13500

48

References

Clark, F. D., and A. B. Lorenzoni. 1997. *Applied cost engineering*, 3rd ed. New York: Marcel Dekker.

Cotterell, M., and R. Hughes. 2002. *Software project management*. 3rd ed. London: McGraw-Hill.

Elmaghraby, S. E. 1977. *Activity networks*. New York: Wiley.

Granli, O., P. W. Hetland, and A. Rolstadås. 1986. *Applied project management*. Trondheim, Norway: Tapir Forlag.

Guthrie, W. 1977. *Managing capital expenditures for construction projects*. Carlsbad, CA: Craftsman Book Company.

Jessen, S. A. 2002. *Business by projects*. Oslo: Universitetsforlaget, 2002.

Lockyer, K. 1984. *Critical path analysis and other project network techniques*. 4th ed. London: Pitman.

Morris, P. W. G. 1994. *The management of projects*. London: Thomas Telford.

Turner, J. R. 1999. *The handbook of project-based management*. New York: McGraw-Hill

Wysocki, K. W., R. Beck, and D. B. Crane. 1995. *Effective project management: How to plan, manage, and deliver projects on time and within budget*. New York: Wiley.

CHAPTER THREE

CRITICAL CHAIN PROJECT MANAGEMENT

Lawrence P. Leach

Critical chain project management (CCPM) is a relatively new entry to the Project Management Body of Knowledge (PMBOK), first reaching the broad project audience with a presentation I gave at the Project Management Institute (PMI) international conference in Long Beach, California, in 1998. This followed the 1997 publication of the book *Critical Chain* (1987) by Eliyahu M. Goldratt, best known for his much earlier production management classic *The Goal* (1984). *Critical Chain Project Management* (Leach, 2000) in February of 2000 integrated the schedule preparation and management method of Goldratt's critical chain approach with the rest of PMI's PMBOK.

The excitement about critical chain stems from it being the first *new thing* to project planning since CPM (critical path method), PERT (program evaluation and review technique), and Monte Carlo simulations came on the scene in the 1960s. It also results from the reports by some early adopters of CCPM quoting statistics like the following:

- Project success rates (triple constraint): Up from < 10 percent to near 100 percent.
- Projects complete in half the previous time, or less.
- Much less stress on project managers and team members.

This early excitement was tempered in the PMI community by charges that CCPM really is nothing new; that everything it includes was already available in the PMI's PMBOK®. As companies beyond the early adopters sought the benefits promised by CCPM, some found that it required changes in management behavior they were not ready for. Consequently, a growing number of CCPM implementations did not get the rapid benefits promised by the early adopters. CCPM is now becoming a standard tool for project planning and control. This chapter explains why.

Any Project Worth Doing Is Worth Doing Fast!

Projects we are concerned with all have a purpose that can be expressed in terms of return on investment. The return need not be monetary; for example, it may be lives saved by a humanitarian project. Project investment can usually be expressed in monetary terms; but even if you choose to express it in terms of staff-hours or some other measure of resource consumption, the following applies. Most projects provide no return (or at least very little) until they are done. Thus, return on investment, no matter what the units of measure, is negative until the project completes, and then begins to pay back and hopefully eventually becomes positive. Because the investment is roughly the same whether you complete a project slowly or fast, the return on investment accelerates in time if the project completes sooner, with no additional investment. (Worse, projects that take longer usually cost more for the same final result.) Further, if the project is a new product development, the overall profitability could be impacted by hundreds of percent by being first to market. Therefore, any project worth doing is worth doing fast.

The basic idea underlying CCPM derives from the theory of constraints (TOC), developed by Dr. Eliyahu Goldratt (Dettmer, 1997). There is no consensus on the basic statement of this theory. It starts as a system theory, recognizing that people build systems to achieve a goal. Usually businesses can best measure progress toward the goal as throughput of the business. TOC asserts that goal achievement for any system is limited by a constraint. It then follows that if you do not know what the constraint is, most of the work you might do to improve the system will improve things that are not the constraint and thus lead to no improvement of the system in terms of its goal (which, for profit-making companies, is to make money now and in the future). Goldratt had great success taking this simple idea into the world of production with five focusing steps for system improvement:

- *Identify* the constraint.
- *Exploit* the constraint (i.e., do whatever is necessary to ensure the constraint works at full capacity on quality input to produce quality output and pass on the work result as soon as possible).
- *Subordinate* everything else to the constraint (i.e., eliminate interferences with exploiting the constraint to achieve system throughput). Often this includes eliminating subsystem efficiency measures and policies.
- *Elevate* the constraint (i.e., get more of the constraint, be it machines or people).
- Do not let *inertia* keep you from doing the cycle again (as a new constraint always arises).

TOC accounts for statistical fluctuations in process step time and the dependent flow of material through process steps in a production line. Both of these factors apply to projects as well.

Single Project Critical Chain Project Management

Identify the Constraint

Applying TOC to project management immediately confronts one with the triple constraint (scope, time, budget). Although called the triple constraint, the idea of limited resources always lurks nearby. TOC cannot abide a triple constraint, much less a quadruple constraint. The first step is to identify the constraint of most projects. Keep in mind the opening remarks; it is the constraint to the goal, and the constraint to getting more of the goal sooner, that is of interest.

One finds much discussion of the critical path in the project literature . . . a concept we need not redefine in this chapter. Figure 3.1 illustrates a sample critical path schedule as a point of departure to illustrate the single project constraint, and how critical chain differs from the critical path method (CPM).

Two things about the critical path are interesting from a TOC perspective. First, most projects have finite resources. Second, activity performance times always show statistical fluctuations. CPM does not take either of these factors into account. The project literature suggests that you should resource-level your CPM plan. Figure 3.1 illustrates a plan without resource leveling, assuming one resource of each type shown on the plan. (Many project managers do not even identify the resources needed in the project plan. Identifying the required resources is called "resource loading" the plan.) Note that tasks 3, 9, and 14, which are planned in parallel in Figure 3.1, all require an engineer. If there is only one engineer and these tasks assume the engineer is dedicated to the task, these tasks must overrun by at least a factor of three.

Try it and you will find that the overall project duration is significantly longer than the sum of the critical path durations. Figure 3.2 illustrates the resource-leveled version of Figure 3.1. Note that the plan now only requires one resource of each type at a given time. The plan is substantially longer. Thus, the critical path is not the constraint to getting the project done sooner. As such, it is questionable if it is the critical path. Further, as illustrated in Figure 3.2 (the space between task 3 and 14), the critical path usually has gaps in it after resource leveling—that is, apparent float. The PMBOK® Guide (PMI, 2000) defines the critical path as the path with zero float. Thus, after resource leveling, no path may qualify as the critical path.

Critical chain corrects this logic problem by considering equally both the resource constraint and the activity logic when defining the critical chain. The critical chain is the longest path through the project . . . it is the real resource-leveled critical path. It jumps the logic chains whenever a resource contention requires it to do so. Except for rare exceptions, the critical chain is not the same sequence of activities as the critical path. Figure 3.3 illustrates the critical chain for the example project. This chapter addresses the changed activity duration and the buffers.

Exploit the Constraint

The second TOC focusing step is to exploit the constraint, which is the critical chain of the project. Exploit means to get more out of the system with no more input. For the reasons discussed previously, this means complete projects quicker. Goldratt relied on his production experience to suggest a direction for the solution to this step, taking into account the statistical fluctuations in project activity times and the dependent nature of the project logic network.

FIGURE 3.1. EXAMPLE CRITICAL PATH SCHEDULE.

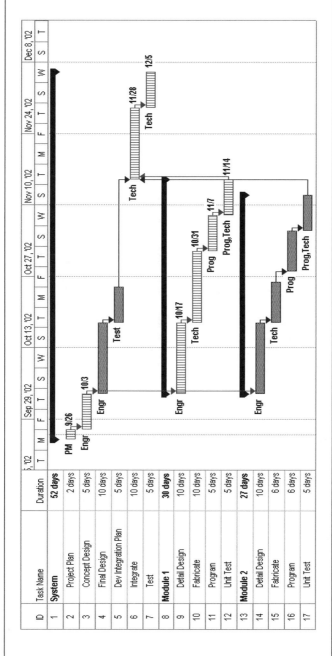

FIGURE 3.2. RESOURCE-LEVELED CRITICAL PATH.

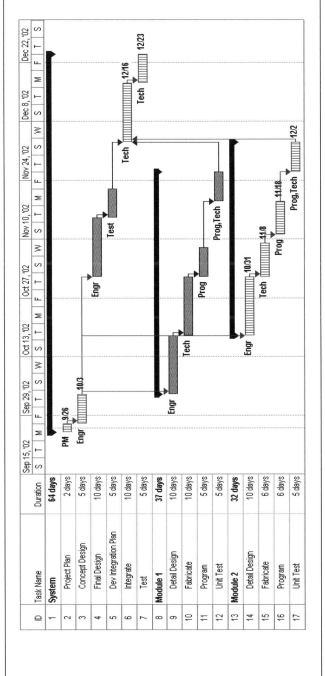

FIGURE 3.3. CRITICAL CHAIN FOR EXAMPLE PROJECT.

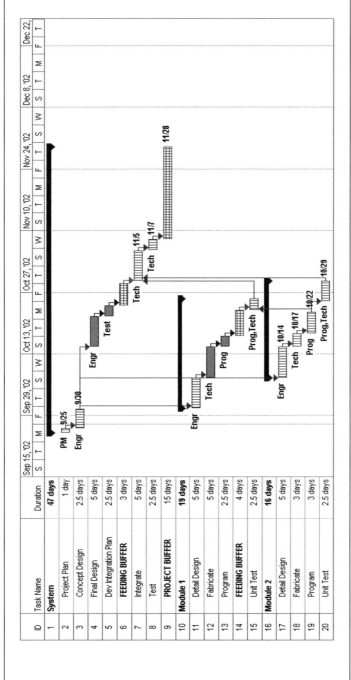

Mean Estimates. Most project work is known to have a completion time probability density function something like that shown in Figure 3.4. Functions that approximate this shape are the mainstay of probabilistic project planning methods such as PERT and Monte Carlo simulation.

There are actually at least two distributions that have characteristics similar to Figure 3.4. One results from simply doing the same activity over and over and measuring how long it takes. That is a measure of the actual variation in activity performance duration. Numbers that you find in databases for estimating activity performance—for instance, the probability distribution of bricks laid per hour—show such a distribution. Many project activities seem unique or first of a kind. One of the primary tools of project management is to use the work breakdown structure to decompose such activities down until you have activities on which you have an experience base to make an estimate.

The second contribution important to project schedules is the variation in estimation. Ultimately, project schedulers are interested in the comparison of actual duration to the estimated duration. If different estimating techniques are used, one can get different results from the same data. For example, even the most basic construction activities allow for adjustments for worker productivity by region or for whether one is working on ground level or on a scaffold.

Consider Figure 3.4 to understand another point about activity estimates. The probability associated with any point on the x-axis—that is, with any specific duration estimate—

FIGURE 3.4. ACTIVITY TIME PROBABILITY DISTRIBUTION.

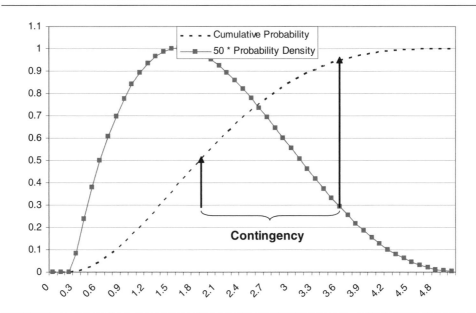

is exactly the same. It is zero. In a conventional schedule with single deterministic schedule duration estimates, all of the individual activity start and stop dates (and with CPM usually both early and late start and stop dates) thus have a probability of zero . . . they are a complete fiction. Yet people get driven to those specific dates.

One only gets a probability by integrating between two points. Usually when one gives a single point estimate, the meaning is "within that time or less"; in other words, the associated probability is the integration from zero up to that duration. Figure 3.4 shows that as the dashed curve. This is why statisticians insist that point number estimates, without some indication of the associated uncertainty, are meaningless.

Project managers face the problem of what to do about such distributions. The PMBOK® Guide encourages estimating the uncertainty for all project estimates. Most current software tools include the PERT capability. Despite these facts, most project schedules do not address uncertainty. Most schedules use a single activity duration estimate and focus on calculated start and finish dates for each activity.

PERT attempts to model the curve of Figure 3.4 by making three estimates (time or cost): an optimistic, most likely, and a pessimistic estimate. It uses these estimates and an assumed distribution (usually a version of the beta distribution with specific parameters, or a triangular distribution, which never exists in reality) to estimate the mean value and variance for each activity along the critical path. Critical chain uses essentially the same approach, but often with only two estimates, and sometimes with no explicit estimate of the individual activity variation. PERT usually limits this thinking to the critical path, while critical chain applies the same thinking to all activities in the schedule. Monte Carlo simulations attempt to do the same thing and usually allow many more degrees of freedom on the statistical distributions applied. This is an area where the claim that critical chain is "nothing new" is valid . . . people have understood the element of uncertainty from the beginning of project management and have been making attempts to use the knowledge. CCPM takes the same knowledge and builds a different approach on how to use it through integrating other elements of the process, such as measurement and control.

While some project literature (correctly) urges using mean times for the deterministic activity duration estimate, the culture that persists around the world does not support using mean estimates. It is correct to use the mean because individual mean estimates along a path do add linearly, whereas other central tendency statistics (e.g., mode, median) do not. But everywhere you look, people are judged as good if they get done sooner than an activity estimate and bad if they are late. Thus, people are behaviorally reinforced to make estimates that are much higher probability than the mean, perhaps as high as a cumulative probability of 90 to 95 percent. For most activity distributions like Figure 3.4, the estimated duration must therefore be two to three times the mean.

I find much of the project literature emotionalizes the reality of statistical fluctuations. They talk about activity time "padding" as if there were some absolute reference to pad against (the mean, mode, median?). Some suggest adding an activity at the end of the critical path to account for these fluctuations . . . but quickly caution to call it something that will not draw attention to it, as though if you do, management will take it away. This thinking illustrates poor understanding of statistical fluctuations. Such poor understanding leads to mismanagement of uncertainty. A key aspect of CCPM is to recognize and utilize the fact of statistical fluctuations. To do that, you need a tool.

Project Buffer. The project buffer is the primary tool to manage statistical fluctuations. Using the mean duration for all project activities, the probability distribution of the end of chain of activities will approach a normal distribution due to the central limit theorem of statistics. This means there is only a 50 percent chance of completing the project at that time or less. Thus, it is necessary to add the project buffer as an allowance to the end of the critical chain to bring up the probability of delivering the project on time (or earlier) to some desired level; usually > 95 percent. Figure 3.3 illustrates a project buffer as activity 9.

Project Buffer Sizing. Properly sizing the project buffer has been a matter of much discussion. Goldratt initially suggested starting with activity estimates as people had always made them, cutting the individual estimates in half and adding back in a project buffer equal to one half of the resulting critical chain. Although the subject of much criticism, this method has shown its merits many times.

I believe that the best project companies will eventually come to use control charts to size project buffers. Tracking a control chart of the difference between scheduled and actual completion time for projects reveals the amount of adjustment necessary to the mean project duration to achieve any desired level of predictability.

In the meantime, many people have suggested a method that is similar to PERT: sizing the buffer as the square root of the sum of the squares (SSQ) of the differences between low-risk and most-likely or optimistic activity duration estimates (Newbold, 1998, p. 94). This method has the same advantages and disadvantages of the PERT approach. Leach (2003) suggests modifying this method to account for sources of bias in project performance to plan, with a minimum project buffer limit of 25 percent of the length of the critical chain. The sources of bias (that is, a tendency for projects to overrun the schedule and not underrun) include the following:

- Oversights (necessary activities not included in the plan).
- Errors.
- Merging project activity chains (more than five or six, or nearly equal in length and uncertainty).
- Queuing effect of multiple projects demanding the same resource.
- Overconfidence as to the variation of activity performance. This is institutionalized in PERT, where there is an inherent assumption that the range people estimate is plus-or-minus three sigma. Analytical evidence demonstrates people's estimates usually range more on the order of plus-or-minus one to two sigma (Kahneman 1982).
- Multitasking.
- Student syndrome: Delaying the start of activities until the need seems really urgent, such as students do with term papers.
- Date-driven behavior: Failure to turn in the results of activities that complete early; spending the time "polishing the apple" or sitting in an out box until the due date.
- Failure to report rework: Passing on work with known defects to meet the scheduled due date. The rework gets discovered later, where it has a much larger impact on completing the project.

The effects of the first two items on this list are evident, and this chapter also discusses merging and multitasking.

One may use Monte Carlo simulations to size the project buffer, but they are subject to most of the preceding biases that the PERT and SSQ approach overlook. An exception is that some Monte Carlo schedule analysis tools may take path merging into account.

Subordinate to the Constraint

Feeding Buffers. Most project networks have multiple parallel chains. All chains must tie in by the end of the project. You should always have a milestone at the end of the project schedule for project completion. Many chains tie in to the critical chain before the end of the project, usually into assembly, test, or transition to operation activities. Any of these merging chains can make the project critical chain late, as the successor activities require all input activities to be complete before they can start. In CCPM, these merging chains are called *feeding chains*, as they feed the critical chain.

Conventional project management seeks to use float to manage feeding chains. *Float* is the result of a backward pass calculation on the network and reveals how much later than the start of the project activities can start and still not make the project late. By definition, the critical path or critical chain has zero float. The assertion is that the float allows for uncertainty in the chains that feed the critical path.

Unfortunately, float is a very poor tool to account for uncertainty in merging chains. Float has nothing to do with the uncertainty of the duration of the activities in the feeding chain. Float results from only network logic construction. In general, longer feeding chains have less float. Thus, as Figure 3.1 illustrates, the amount of float available is inversely proportionate to the amount needed. (Compare the float for activity 5 to that for activity 17.) A chain just short of the critical chain, such as the lower one in Figure 3.1, has nearly zero float. Yet, as the next longest chain in the project, it needs the most float.

CCPM inserts feeding buffers at the point feeding chains join the critical chain. They connect the last activity on the feeding chain to the successor on the critical chain. Figure 3.3 illustrates two feeding buffers for the sample project. The scheduler sizes feeding buffers based on the uncertainty of the duration of the activities in the feeding chain, applying the same method used to size the project buffer. In addition to providing assurance that the successor critical chain activity will have the necessary input from its predecessor on the critical chain and on the feeding chain, feeding buffers also enable measuring progress on the feeding chains.

A simple illustration serves to show the importance of feeding buffers to the overall schedule. If each chain of activities has a reasonable number of activities, and if mean estimates are used, the probability distribution for completing the chain as scheduled would have a 50 percent probability of completing at the merge point. Thus, the chances of having both are only 25 percent, the product of 0.5×0.5. If a third path joins at the same point, the probability of having all three reduces to just 12.5 percent. And so on. Thus, it is difficult to protect a highly parallel schedule. Analysis demonstrates that sizing the feeding buffers using the same method as recommended for project buffers is effective for a modest number of feeding chains: up to five or six. More merging chains require larger feeding buffers, or an addition to the project buffer.

Late-Start Feeding Chains. Most project software defaults to early-start scheduling. This may be for two reasons:

1. The belief that the float thus introduced helps manage project activity time variation
2. No alternative but late-start

This in turn can create several problems for project execution:

1. The project manager is confronted with a bewildering array of activity chains to start while at the same time needing to align all of the project stakeholders.
2. Early-start can cause resource contention not otherwise present, leading to bad multi-tasking (see *Resource Decisions* coming up in the chapter for a more thorough description of bad multitasking.)
3. If one is using earned value (EV) against an early-start schedule, the project will look behind from the beginning, even if it is doing fine. This can cause the project manager to take unnecessary actions, leading to increased variation in project performance.

CCPM recommends late-starting feeding chains. Keep in mind that each feeding chain has a feeding buffer activity where it joins the critical chain, meaning that the start of the chains will not be delayed as much as in CPM or PERT late-finish schedules. It is applying the just-in-time principle to project performance, while using the feeding buffer to account for the reality of activity duration variation.

Earned Value and CCPM. One may apply earned value (EV) to CCPM projects. Because the schedule variance (SV) and Schedule Performance Index (SPI) do not reflect the critical chain (or path), and because they can provide misleading information on actual project status and even encourage inappropriate behavior (e.g., performing expensive noncritical tasks before inexpensive critical chain tasks), I recommend not using SV or SPI for project control or forecasting. Buffer management provides the schedule forecasting and control functions for CCPM projects.

When applying earned value to CCPM projects, you must consider the statistical nature of the estimates: The task estimates are the mean values, and the buffers statistically sum the variances. When providing a high-probability estimate of project cost and schedule completion, project completion is at the end of the project buffer. The baseline budget at completion (BAC) must include an allowance for the cost of working into the project buffer. This cost allowance is commonly called a *contingency* or *management reserve*, but is usually not integrated with the schedule impact, as it must be in CCPM. The project baseline completion date and the baseline BAC for the project must include these buffers. The planned value (PV) for the project extends to the end of the buffers in both time and money. Figure 3.5 illustrates this compared to the standard project S curve (i.e., the accumulated project

FIGURE 3.5. CCPM EARNED VALUE.

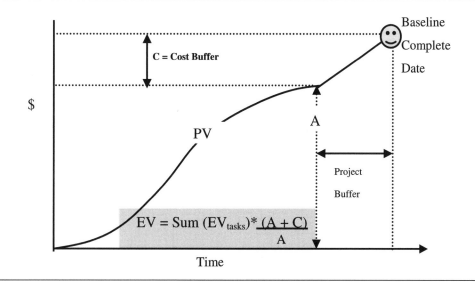

progress and cost vs. time first raising slowly, then rapidly, and then slowly to final project completion).

The sum of the activity estimates is the mean time and cost to complete the project. This must be adjusted to predict how the project will complete relative to the baseline. This is best accomplished by setting EV* = EV (activities) × R, where R = ratio of the BAC* (with contingency) to the BAC = sum of the activity (mean) estimates. With this adjustment, other EV calculations will work normally—for instance, EAC = BAC/CPI, where AC = actual cost, CPI = Cost Performance Index, and CPI = EV*/AC.

For example, assume a project where the sum of the estimated tasks (BAC) is $100K and the cost buffer = $20K. Thus, BAC* = $120K. Halfway into the project (i.e., half the tasks completed), EV = $50K. Let's assume that the project is right on cost: AC = $50K. Then, EV* = 50 * (120/100) = 60. Then CPI = 60/50 = 1.2. Thus, EAC = 120/1.2 = $100K.

Buffer Management

Buffer management provides the CCPM cost and schedule control system. (CCPM projects need effective quality management, just like any project.) Buffer management starts with weekly updates of the schedule by asking all activity performers, "How many working days until you will be done?" This forecast of activity completion is used to project the completion of the project activity chains. Keeping the feeding and project buffers fixed in time allows determining the penetration, incursion, or use of the buffer. Incursion into the project buffer forecasts project completion. Figure 3.6 illustrates tracking buffer incursion.

FIGURE 3.6. TRACKING BUFFER INCURSION VS. TIME.

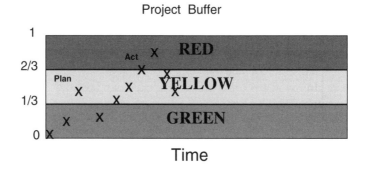

The specific question "How many working days until you will be done?" is important. Experience demonstrates that estimates of percent complete tend to underestimate the time necessary to complete an activity. Causing activity performers to think about what remains to be done, rather than what they have completed, leads to more realistic estimates.

The entire project team uses buffer management to make decisions about the project. The two primary decisions are those made by resources, who must decide which activity to work on next, and by the project manager, who must decide when to take action on the project to keep it in control.

Resource Decisions. A cardinal rule of CCPM is that all resources should work on only one project activity at a time and turn in their result as soon as it is completed. This is called roadrunner or relay-runner activity performance. The reason is that dilution of work time amongst different project activities makes them all take longer and ultimately makes all projects take longer. This is what CCPM calls bad multitasking. An insidious aspect of bad multitasking is that if one keeps a database of activity durations in a multitasking environment, the database will reflect this poor practice. This combined with current practice in many organizations to turn in work when due, but not earlier, leads to a self-fulfilling prophecy of protracted activity duration.

To avoid bad multitasking, each project resource must have a tool that specifies which activity to work on next. The buffer report provides this tool. Each resource should work on the activities in their queue (following the project logic first, of course), in accordance with the following run rules:

1. Critical chain activities take priority over non-critical chain activities.
2. If two activities are both on or both off the critical chain, the resource should work on the one with the greatest relative buffer penetration.
3. Nonproject work is lower priority than project work.

Project resources like to know when activities are coming up to be worked. Some suggest using resource flags or buffers along the critical chain. The resource buffer does not consume time in the project; it is an "alarm clock" lead time to let resources know when a critical chain activity is coming up to be worked—usually with a fixed lead time. Most critical chain implementations do not require this added formality. Project status meetings and communication of project schedules often fulfill this function.

Some resources will have difficulty with not knowing a fixed date to start and complete activities. In reality, they did not have fixed dates before, only estimates. In time, they will come to understand this. Where you must use fixed dates, be sure to precede them with an appropriately sized buffer.

Project Manager Decisions. The project manager can use buffer reporting as a control chart to decide when to take action on the project. As described in two *PM Network* articles (Leach, 2001), it is necessary to discriminate between common-cause (i.e., within the natural variation of the process that produces the result, in this case an actual project task duration) and special-cause variation (i.e., an identifiable cause of variation beyond the natural process limits) before taking action on a project. Mistaking one for the other will increase overall project variation. Control charts set limits on variables such that variation within the control limits represents common-cause variation, and should not be acted upon. Variation outside these limits is special-cause variation, and calls for action by the project manager to bring the project back into control.

Figure 3.6 illustrates a buffer control chart. Limits are preset for planning to take action and for the initiation of action. Variation below the planning threshold requires no action by the project manager. Variation beyond the normal range, but not yet to the trigger point, signals the project manager to develop mitigation plans. Variation beyond the action trigger causes the preplanned action to take place.

One may use buffer management for all of the buffers in the project. In practice, feeding buffers are often used up very rapidly, and tracking on a chart is frequently not necessary. Some suggest that the buffer planning and action thresholds should vary over the planned duration of the project, with tighter limits at the project beginning. I have not found that extra sophistication to have significant value, nor have I found it to be detrimental.

What Is Different about CCPM?

The schedule is different from most CPM plans.

- All CCPM schedules are resource-loaded and -leveled.
- The critical chain jumps logic chains where necessary.
- Mean activity durations are used.
- Project and feeding buffers are used.
- Project start times are sequenced to project priority for access to an organization drum resource.

- A capacity constraint buffer delays project start times.

Task performance differs from many projects.

- Bad multitasking is eliminated.
- Activities do not have start and stop dates; they are performed as fast as possible in sequence as the resource is available.
- Early activity completions are passed on as soon as possible to the nest activity.

Measurement and control is different.

- Status asks "how many days to complete?" working activities.
- The buffer report forecasts project completion.
- The project manager uses buffer thresholds to take pro-active project action.
- Resources and resource managers use the buffer report to decide which activity to focus on next.

Multiple-Project Critical Chain

CCPM defines a multiproject environment as one in which resources are shared across projects. Informal surveys I have conducted confirm that nearly all project environments qualify. The measure for the multiproject environment is completing the most projects in the shortest possible time to support the business goal. Applying the TOC five focusing steps identifies a different constraint than the critical chain of a single project. One of the resources shared across all of the projects controls the overall throughput of project results. CCPM calls this resource the "drum," with the image of the drummer on an ancient galleon setting the pace for the whole organization.

Figure 3.7 illustrates why it is so important to identify and control using the constraining resource. The upper three projects share represent multitasking of the resource supply over the three projects. In the lower three projects, the resources are sequenced, thus reducing the duration of each activity proportionately.

However, the upper case could represent one person shifting back and forth between the three projects, or three resources, one on each project. Usually having more than one resource helps lighten the load—that is, two people get the work done in less than half the time of one.

I use these simple project constructs in a training simulation, and the result is always the same. If one performs like most organizations do and starts two or more projects at the same time (e.g., the first of the fiscal year), it always takes much longer to complete each project than if the projects are staggered to enable the drum (i.e., the shared resource used most across all projects: the system constraint) resource to focus on one project at a time. More remarkably, both projects complete sooner than either project did in the multitasking case. In real organizations with dozens of projects, the results are much more impressive.

FIGURE 3.7. COMPARISON OF MULTITASKING AND PROJECT SEQUENCING.

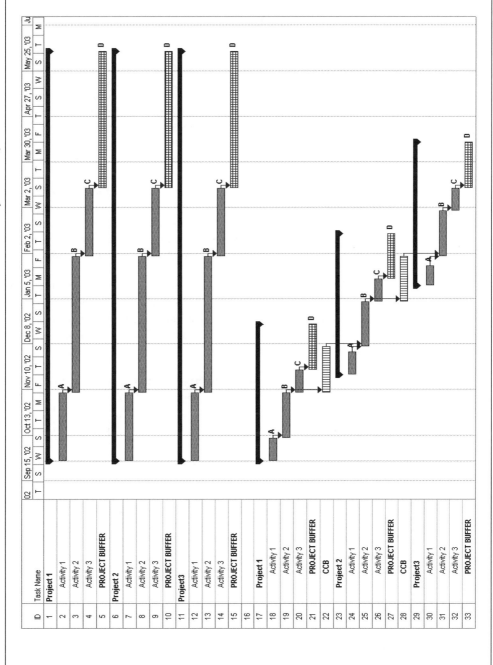

Note that in the lower case, even the last project finishes before any of the projects in the upper case.

Figure 3.7 is conservative, as it assumes no switching loss for multitasking in the upper three projects. Research indicates the multitasking switching loss can be on the order of 40 percent. Multitasking also does not allow for the learning curve to help later projects.

Identify the Constraint. The first focusing step is to identify the constraint—the drum resource. Our purpose in identifying the constraint is to use the information to sequence the start of projects. It has nothing to do with how important the resource is. However, if a resource can be increased relatively quickly and inexpensively, it is not a good choice to remain as your company project constraint. The value of the constraint far outweighs its cost, as it limits your entire project throughput. So it should also be a resource that is used relatively heavily by your projects.

Although identifying the drum is usually not difficult; it can sometimes cause frustration for your project managers. Some prefer to take an analytical approach and add up the resource demands from all of the extant project plans. This usually does not work because resource identification is not sufficiently precise to do so, and the set of resource loaded project plans is often a poor reflection of the actual projects that the organization will perform.

Selection of the wrong resource as the drum is not a fatal mistake. You will get information to correct the error. More importantly, selection of any resource and using it to stagger the start of projects will improve your organization performance. Selecting the right resource will improve it the most, but often the difference of staggering projects to any resource far outweighs the importance of having the right constraint to start with.

Project Priority. The next step is to prioritize all projects that will be performed. The priority is for access to the drum resource and for scheduling the start of projects. It has little to do with the absolute value of the project, although if your information is good enough to do so, you should place higher priority on the projects that deliver the greatest throughput per use of the drum resource.

Project prioritization must become an ongoing process. All new projects must be placed in the priority list before they are scheduled. You can adopt any priority rules you wish, but you must stick to them on an ongoing basis.

In parallel with setting the project priority, plan each project using the planning tool to create a resource loaded critical chain schedule, just as in the single project case. Each project should assume access to the maximum amount of each resource that it can use effectively, as if it was the only project the company was doing. Each project should assume 100 percent dedication of the resources over the duration of the activity.

Subordinate to the Constraint. Resource B is the constraint (drum) in the Figure 3.7 case. A final step necessary to get the start and completion of the projects is to place a properly sized capacity constraint buffer (CCB) between the use of the drum resource in one project and its use in the next. Keep in mind that the drum schedule uses mean activity times. Figure 3.7 shows the placement of the CCB between the lower projects. It does not actually appear in a project schedule, but it does set the start date for the schedule by establishing a start-no-earlier-than constraint on the first drum using activity in the downstream project.

The manager of the identified drum resource can then take all of the project plans and their priorities and create a schedule for the work of the drum resource. The schedule for the drum resource on each project will then determine the start and completion date of each project. (This discussion can be an oversimplification if projects use the drum resource for many tasks. Describing the process for those cases goes beyond the scope of this essay, but the illustration here adequately describes the function of the process.)

Multiple-project CCPM does not resolve all resource contentions across all projects. That is one of the great simplifications provided by the TOC approach. Experience demonstrates it is not possible to maintain information integrity across all projects on a timescale comparable with the rate at which things change. Since all non-drum resources have excess capacity, on a statistical basis they cannot control project output. Because of statistical fluctuations, there will be times that the demand for nonconstraint resources exceeds capacity. The buffer management rules tell the resources the sequence of tasks to work on. The buffers absorb the impact of the delay of some tasks.

Multiple-Project Buffer Management. Multiple-project buffer management adds one rule to the resource decisions. Project priority comes first. Then, the remaining buffer management rules apply.

Organization-Wide Implementation of CCPM

Any individual can productively use the principles of TOC and CCPM. For example, you can eliminate bad multitasking in everything you do. You can use the five focusing steps at home and at work to focus your energy where it matters most. Most project managers can use mean estimates, buffers, the critical chain instead of CPM, CCPM status determination and reporting, and buffer-threshold-based decision making without disturbing the rest of your company. But if you are in a multiproject environment (as most of us are), to really supercharge your company's project results, you are going to have to lead a company-wide change.

Geoffrey Moore wrote two books about bringing a new product into the mainstream. You can think of CCPM (or any significant new management process) as a new product, and the managers and employees of your company as the market. *Crossing the Chasm* (Moore, 1991) describes how the original new product adoption life cycle traces through a series of customers, starting with Innovators and proceeding through Early Adopters, Early Majority, Late Majority, and Laggards. He added the concept of the chasm between the Early Adopters and Early Majority. Most new products die in this chasm. Most new management processes (which some call fads; e.g., excellence, TQM, just-in-time, teams, reengineering) die in this chasm as well.

To succeed, you must plan and manage your implementation like a new product introduction. Although new products such as TQM worked very well for the early adopters, they fell into the chasm in later companies for a variety of reasons. Moore suggests, "The winning strategy does not just change as we move from stage to stage, it actually reverses the prior strategy . . . [it is] the need to abandon each one in succession and embrace its opposite that proves challenging.

Usually the ones interested in CCPM at first are the techies . . . new project managers, schedulers, project control people, and in some cases software people. Your best project managers may not feel they need it, because they have learned how to succeed against all odds and aren't necessarily interested in making it easier on their internal competitors. Early adopters are visionaries who will look to the promised benefits and give it a try. You may find a senior management sponsor willing to take on this role. If so, you can make it to the chasm. To cross it, you have to bring along the Early Majority. They are the pragmatists. These "show me" people are generally more interested in evolution than revolution. Moore characterizes them as people who "look to adopt innovations only after a proven track record or useful productivity improvement, including strong references from people they trust." This means you are going to have to run some credible pilot projects in your company to convince these people and bring them along.

Before CCPM becomes "the way we do projects around here," you need to bring the Late Majority on board. They are the real conservatives. They are pessimists about anything new and are only going to change what they are doing under duress. The key to getting them to follow is to make it simply easier to do it the new way, and perhaps to close the door to doing it the old way.

You can safely almost forget about the Laggards. They will not change. They are skeptics, and not worth the investment it will take to bring them along. You simply have to keep them from blocking your way.

Implementing CCPM requires changing certain behaviors. For example, in many companies the ability to multitask is considered a virtue. When you are deploying CCPM, it is a vice for any project resources to multitask while working on a project activity. Usually multitasking behavior is a result of feedback from management that encourages it and often due to the inability of management teams to agree on a company priority. Functions tend to suboptimize and do what is best for their function, instead of what is best is for the company. Frequently this is at least in part because they have no way of knowing what is best for the company. Tables 3.1 through 3.6 list potential behavior changes that you may face. Not all companies reflect the present behavior listed so you may not have to address all of the behaviors as a change from a contrary behavior. You do have to succeed to create the CCPM behaviors.

Because these behaviors have been in place a long time, it will be difficult to change them. It usually requires focusing on changes at the individual, social group, and management level to achieve the full benefits of CCPM. Hellriegel et al. (1998) poses an organization behavior framework that addresses individual processes, group and interpersonal processes, organizational processes, and organizational issues. They clarify that you must consider this model in light of environmental influences that affect behavior in each element of the model. They note that "the management of change involves adapting an organization to the demands of the environment and modifying the actual behavior of employees."

The following sections outline the actual behavior of employees' changes that many companies face. Each company has to develop its own plan for how to cause the change.

As a common saying goes, "If you always do what you've what you've always done, you always get what you always got." You can't expect to double throughput without changing what you do.

Some people adapt to change better than others—they are the Early Adopters and your initial customers. Most people follow the crowd; your strategy is to gradually enlist

TABLE 3.1. BEHAVIOR CHANGES SENIOR MANAGEMENT MUST MAKE.

Change	Present Behavior	Future Behavior
Only commit to feasible delivery dates.	Sometimes commit to arbitrary delivery dates: determined without consideration of system capability to deliver.	Only commit to delivery dates with a critical chain plan and (if multiple projects), after sequencing through the drum schedule.
Eliminate interruptions.	Insert special requests to the system with no assessment of system capability to respond. Sometimes place demands for routine administrative work above project work (e.g., salary reviews).	Prioritize all requests using buffer report.
Set project priority.	No clear project priority, or changing project priorities,	Set project priorities; including the priority of new projects relative to ongoing projects.
Select drum resource.	No consideration of system constraint.	Select the drum resource to be used for sequencing the start of projects and creating the drum schedule.
Select drum manager and approve project sequencing.	Start each project independently as funding is available.	Drum manager creates drum schedule. Senior management approves. Project managers schedule projects to the drum.
Project Status.	Looking over shoulders.	Buffer report

TABLE 3.2. BEHAVIOR CHANGES RESOURCE MANAGERS MUST MAKE.

Change	Present Behavior	Future Behavior
Resource priority.	Assign resources on a first come, first served priority; or attempt to meet all needs by multitasking.	Assign resources using the buffer report.
Resource planning.	Plan resources by name and activity.	Plan resources by type, and assign to activities as they come up using the buffer report priority.
Early completion.	Turn in activities on due date.	Turn in activities as soon as they are complete.
Eliminate multitasking.	Ensure resource efficiency by assigning to multiple activities at the same time.	Ensure resource effectiveness by eliminating bad multitasking.
Resource buffers.	Resources planned far ahead and not available when needed.	Use resource buffers and buffer report to dynamically assign resources to activities.

TABLE 3.3. BEHAVIOR CHANGES PROJECT MANAGERS MUST MAKE.

Change	Present Behavior	Future Behavior
Mean activity duration estimates.	Project managers send a message that they expect due dates to be met.	Project managers first get low-risk activity duration, and then get "average" duration; using activity uncertainty to size buffers.
Roadrunner/relay racer activity performance.	Provide start and finish dates for each activity, and monitor progress to finish dates.	Provide start dates only for chains of activities, and completion date only on the project buffer.
Feedback on activity duration overruns.	Management provides negative feedback when activities overrun due dates.	Management provides positive feedback and help if resources perform to roadrunner paradigm.
Project status.	Varies. Often use earned value as the schedule measure.	Buffer report (including a cost buffer).
Project changes.	Varies. Often submitted to minimize minor variances.	When triggered by buffer report.
Response to management demands for shorter schedule.	Arbitrary activity duration cuts.	Add resources or make process changes to get a feasible and immune schedule.
Early start.	Start activities as early as possible.	Start activity chains as late as possible, buffered by feeding buffers.
Sequence projects.	Start project as soon as funding is available.	Schedule project start using drum schedule.
Assign resources dynamically according to critical chain priority and buffer report.	Get resources as soon as project funding is available, and hold resources until they can't possibly be used anymore on the project.	Get resources only when needed, and release as soon as activity is complete.

TABLE 3.4. BEHAVIOR CHANGES PERFORMING RESOURCES MUST MAKE.

Change	Present Behavior	Future Behavior
Perform 100% on one activity at a time.	Multi-activity to satisfy all management demands.	Eliminate bad multitasking (i.e., multitasking that lengthens project activity duration.)
Turn in early completion.	Turn in on due date.	Turn in as soon as "good enough."
Eliminate (bad) early start.	Start as soon as funding is available.	Start chains as scheduled.
50-50 duration estimates.	Provide protected (low-risk) activity duration estimates.	Provide 50/50 duration estimates.

TABLE 3.5. BEHAVIOR CHANGES SUBCONTRACTORS MUST MAKE.

Change	Present Behavior	Future Behavior
Deliver to lead times.	Deliver to due dates.	Deliver to lead times.
Shorten lead times.	Deliver to due dates.	Shorten lead times.

those groups be demonstrating success with the change leaders. The people you enlist will provide the testimonials to bring the later crowd along.

Moore's second book, *Inside the Tornado* (1995), deals with the phases of product growth after the chasm. In the case of CCPM implementation, crossing the chasm means you have had a few successful pilot projects. Once you have crossed it, you can have unprecedented success bringing this innovation to your company . . . rapid growth that Moore equates to the chaos of a tornado. As the leader of this innovation, it will tax your resources mightily. But if you do not lead it properly, your innovation will fall by the wayside along with the other management fads your company has tried. To quote Moore:

> The market is not yet ready to buy in as a whole. It still has too much invested in the old paradigm and will drag its feet for some time to come. If you try a broad frontal assault now, you will only consume your resources in advance of the real opportunity. Instead, it is time to focus on winning niches of the marketplace, made up of customers who are marginalized under the old paradigm, not well served by it, and who find themselves under pressure to reengineer their businesses to become more competitive.

The market and customers we are addressing are the other project managers and resource managers in your company.

Eliyahu Goldratt (1996) posed a model for organizational change called the "layers of resistance model." This model focuses on change from the perspective of answering concerns

TABLE 3.6. BEHAVIOR CHANGES PROJECT CUSTOMERS SHOULD MAKE.

Change	Present Behavior	Future Behavior
Eliminate project scope changes.	Customers spend little time initially establishing requirements, and then introduce late changes	Establish requirements as part of the project work plan; change as little as possible with formal change control.
Support using project buffer.	Customers interpret contingency as "fat."	Customers understand the need for buffers to reduce project lead time and ensure project success.
Eliminate arbitrary date milestones.	Demand arbitrary date milestones.	Use plan to set milestones.

of individuals. Two of these concerns are obstacles within the organization that might prevent deploying CCPM and potential unintended consequences if the changes succeed. It helps to reduce resistance if you solicit and address these concerns as part of your change plan. I believe that these considerations are necessary, but not sufficient to bring about the behavior changes needed for complete success with CCPM.

Software

Four commercial software packages are currently available for deploying CCPM, and more are expected. Eventually we expect to see it as an option in all schedule software.

It is feasible and reasonable to plan and manage critical chain projects with primitive project planning and control tools. Some very large construction projects ($250 million) did it using colored pieces of paper to represent resources. The following are three software developers currently providing CCPM software.

CCPM+

The latest offering on the market, developed under the direction of the author, CCPM+ provides an add-in to plan and execute critical chain projects. It allows using more features of MS project than other offerings (e.g. task priority and a variety of relationships) and employs an innovative algorithm to help resources determine which task to work on next. More information on CCPM+ is available at http://www.advanced-projects.com/CCPM+.htm.

Concerto

Concerto is an innovative solution to deploying CCPM in a multi-project environment. It employs many features helpful to larger organizations, including a WEB-based interface to update progress and display a variety of task and resource information in useful formats. It is the most expensive tool to implement CCPM. More information is available at www.realization.com/.

Sciforma PS8

Sciforma PS8 is a full-function tool that rivals Microsoft Project in all ways and has the inherent ability to deploy CCPM. It is intuitive and easy for anyone skilled in other mainline project software (e.g., MS Project, Primavera) to adapt to. It is very flexible in reporting and well suited to multiple projects of any size. More information is available at www.sciforma.com/products/ps_suite/ps_suite.htm.

Any critical path scheduling tool can aid planning and controlling critical chain projects. Most such tools have the ability to resource-load and -level the plan. You can put in buffers as dummy activities. You can put in the resource constraint using the activity-linking capability of the software. Although this will become daunting if you have more than about 50 activities in your plan, it provides some control advantages over using off-the-shelf software that usually contains "undocumented features" that can influence your plan. The specific procedures that you must develop depend on the software you use.

Summary

CCPM integrates the critical chain approach to scheduling and schedule control with the rest of the PMBOK (PMI, 2000). It is one step in the process of ongoing improvement to the project delivery system. But it is only that. Projects often go awry because of causes that critical chain does not address at all, such as failure to get and maintain stakeholder alignment, poor scope definition, or ineffective quality assurance or change management. Critical chain only addresses common-cause variation impacts on the project. CCPM adds conventional deterministic project risk management to address special cause variation (Leach, 2001).

When the other PMBOK® necessary conditions are in place, CCPM has the potential to reduce overall project duration for single projects by up to one-half. Overall, project duration reduction in the multiple-project environments can be significantly larger, depending on the amount of bad multitasking the organization engages in. Much of this acceleration comes about because the information provided by critical chain schedules and buffer management enables resources to avoid bad multitasking.

Critical chain enforces the discipline of developing effective schedule logic, resource loading, and resource leveling (Elton, 1998). For many project environments, taking these steps without critical chain would resolve much of the project chaos and improve schedule predictability.

Most importantly, critical chain uses the reality of variation in activity performance duration and uncertainty in estimates. It integrates the knowledge of variation in a simple way that makes it unnecessary for project managers to become experts in simulation or statistics. It connects the method to address uncertainty in developing the schedule with the measurement and control system (buffer management) in a way missed by previous approaches to analyze uncertainty.

References

Dettmer, W. H. 1997. *Goldratt's theory of constraints: A systems approach to continuous improvement.* Milwaukee: ASQC Quality Press.

Elton, J., and J. Roe. 1998. *Bringing discipline to project management. Harvard Business Review* 76 (2, March–April): 78–83.

Goldratt, E. M. 1985. *The goal.* Great Barrington, MA: North River Press.

———. 1996. *My saga to improve production.* New Haven, CT: Avraham Y. Goldratt Institute.

———. 1997. *Critical chain.* Great Barrington, MA: North River Press.

Hellriegel, D., J. W. Slocum, Jr. and R. W. Woodman. 1998. *Organizational behavior.* 8th ed. Cincinnati: South-Western.

Kahneman, D., P. Slovic, and A. Tversky. 1982. *Judgment under uncertainty: Heuristics and biases.* New York: Cambridge University Press.

Leach, L. P. 2000. *Critical chain project management.* Boston: Artech House.

———. 2001. *Putting quality into project management.* Parts I and II *Per Network.* Newtown Square, PA: Project Management Institute.

———. 2003. *Schedule and cost buffer sizing: How to account for the bias between project performance and your model. Project Management Journal* 34 (2, June): 34 ff.

Moore, G. A. 1991. *Crossing the chasm.* New York: HarperCollins.

———. 1995. *Into the tornado.* New York: HarperCollins

Newbold, R. 1998. *Project management in the fast lane.* Boca Raton, FL: St. Lucie Press.

Project Management Institute. 2000. *A guide to the Project Management Body of Knowledge.* Newtown Square, PA: Project Management Institute.

CHAPTER FOUR

PROJECT PERFORMANCE MEASUREMENT

Daniel M. Brandon, Jr.

This chapter concerns project performance both from the classical shorter-term "tactical" perspective and the longer-term "strategic" perspective. Despite many innovations in project management and performance methods, many projects fail; in some industries, such as information technology (IT), most projects still fail. A Standish Group International study found that only 16 percent of all IT projects come in on time and within budget (Cafasso, 1994; Johnson, 1995). Field's study discovered 40 percent of information services (IS) projects were canceled before completion (Field, 1997). The problem is so widespread and typical that many IT professionals accept project failure as inevitable (Cale, 1987; Hildebrand, 1998).

This continued failure in project performance can be attributed to a large degree to lack of long-term successful strategies and accompanying metrics. Earned value analysis (EVA) has proven successful for accurate performance measurement and improvement for time and cost on a single project (within a constant scope and quality). And while time and cost values (and accurate estimates of those values at project completion) are very important, there are other important success criteria for project-based work including employee attitudes, the satisfaction of all stakeholders, and leaving a legacy of relevant "stuff" for future projects. Clear identification of these success criteria (and the underlying critical factors leading to success for each criteria) is essential for overall project success. For longer-term success, a system of methods and metrics has not yet been widely adopted for the project management discipline. The balanced scorecard (BSC) is a strategic long-term management approach that has had considerable success in the last decade. Currently, EVA and BSC are two of the most powerful and successful management tools for measuring and increasing performance. However, the two techniques have not been used together. This chapter explains and illustrates these approaches (EVA, BSC, and critical success factors) and also presents a method to integrate the approaches together to achieve both short-term tactical success and long-term strategic project management success.

Project Performance and Feedback Control Systems

The Project Management Institute defines "process groups" (initiation, planning, performing, controlling, and closing) and "knowledge areas" that are related, as shown in Figure 4.1. Later when the strategic issues in project management are addressed, this "knowledge matrix" may require extension. In this chapter our primary focus is on execution and control; however, while the focus is on just those two process groups, there are key activities within process planning and other groups that are key to both short-term and long-term successful performance improvement.

The basic control process used in project management is the same process used in most engineering and business systems. It is based on the definition and establishment of key measures, then the comparison of those measurements to some desired values or standards to formulate algebraic formulas usually called metrics. If the difference between the measurement and the desired value exceeds some threshold, then corrective action (feedback) of some type is invoked and the degree of corrective action may be a function of the size of the difference (and/or the integration [accumulation] or differentiation [rate] thereof). The measurements may be of process outputs or of the process itself, and the measurement level may be tactical (generally how things are being done) or strategic (generally what things are being done). This is illustrated in Figure 4.2.

What to Measure and Control

Once the decision is made to measure and control performance, the next questions involve "what to measure" and "what to control." Project performance typically involves a trade-off of several dimensions, specifically, what is done (scope and quality) versus the resources used to do the work (time and cost). Scope involves both stated needs (requirements or deliverables) and unstated needs (expectations). Quality involves both built-in quality (internal and/or less readily noticeable aspects) and inspected quality (visible and/or noticeable aspects).

In terms of resources, the consumption of those resources is usually measured. But also important are other factors that might affect the behavior of those resources, including the elements of risk and satisfaction. So a list of things that could possibly be measured include scope (percent complete), time (calendar), cost, risk, quality, and the satisfaction and attitudes of the following:

* Project team, performing organization, and line management
* Benefiting organization ("customer")
* Project sponsors and other "stakeholders" (suppliers, collaborators, etc.)

FIGURE 4.1. PMI PROCESS GROUP AND KNOWLEDGE AREAS.

	Initiation	Planning	Executing	Controlling	Closing
Integration		Project Plan Development	Project Plan Execution	Overall Change Control	
Scope	Initiation	Scope Planning Scope Definition	Scope Verification	Scope Change Control	Scope Verification
Time		Activity Definition Activity Sequencing Activity Duraiton Estimation Schedule Development		Schedule Control	
Cost		Resource Planning Cost Estimating Cost Budgeting		Cost Control	
Quality		Quality Planning	Quality Assurance	Quality Control	
Human Resources		Organizational Planning	Staff Acquisition	Team Development	
Communications		Communications Planning	Information Distribution	Performance Reporting	Administrative Closure
Risk	Risk Identification	Risk Identification Risk Quantification Risk Response Development		Risk Response Control	
Procurement		Procurement Planning Soliciation Planning	Solicitation Source Selection Contract Administration	Contract Administration	Contract Closeout

FIGURE 4.2. THE GENERIC CONTROL CYCLE.

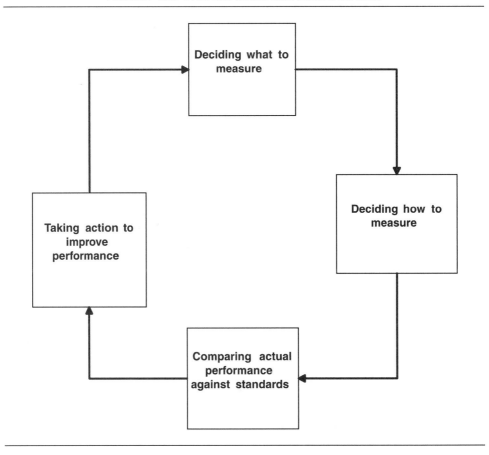

What can and should be measured varies with the type of project and the perspective of the organization managing the project. This is further discussed and quantified later in this chapter under the discussion of success criteria.

Measurement of all relevant variables is important for both management information and also for the specification of "what kind" and "how much" corrective action is necessary. Corrective actions are management prerogatives that are available to a manager based upon the type of organization (functional/hierarchical, projectized, matrix), the position of the project manager, the organization culture, and the governing laws of the state or nation. Examples of corrective action include fast tracking, "crashing," adding additional resources, (people, money, time, etc.), scope reduction, compromising quality, increasing risk, employee/contractor disciplinary actions (negative reinforcement), employee/contractor incentives (positive reinforcement), pep talks and other motivation, and so on. Some corrective actions tend to be more tactical, and some more strategic.

Earned Value Analysis (EVA)—Time and Cost Control within Scope

Traditional project measurement at the tactical level involved only "cost versus budget" and qualitative percent activity complete (Gantt chart). Today, improved quantitative methods are coming into general use such as EVA. PMI now uses the term EVM (earned value management) to describe not only the analysis portion of earned value but the overall management approach of using earned value; both of these topics are included in this discussion. Earned value can be used to more accurately evaluate the performance of a project both in terms of cost deviation and schedule deviation. It also provides a quantitative basis for estimating actual completion time and actual cost at completion. It is one of the most underused cost management tools available to project managers (Fleming, 1994). Refer to Brandon (1998 and 1999) for a more complete discussion of EVA and the effective implementation thereof.

The earned value concept has been around in several forms for many years dating back to types of cost variances defined in the 1950s. In the early 1960s PERT (program evaluation review technique) was extended to include cost variances and the basic concept of earned value was adopted therein. PERT did not survive, but the basic earned value concept did. For many U.S. government contracts, the nature of the contract was such that the government assumed most of the burden for cost overruns. To minimize those overruns, the government was in search of project performance measurements to better control cost. Thus, the earned value concept was a key element in the 1967 DoD (Department of Defense) policy called Cost/Schedule Control Systems Criteria (C/SCSC). Early implementations of C/SCSC met with numerous problems—the most common of which was "overimplementation" because of excessive checklists, data acquisition requirements and other paperwork, specialist acronyms, and overly complicated methods and tools. However, C/SCSC has been refined over the years and is now very effective. The government has accumulated many years of statistical evidence supporting it, and earned value has now met the test of time for nearly three decades on major government projects.

Earned value is basically the value (usually expressed in dollars) of the work accomplished up to a point in time *based upon the planned (or budgeted) value* for that work. The U.S. government's term for earned value is "budgeted cost of work performed" (BCWP). Note that EVA analysis is based upon a predefined scope of work and also a predefined specification for the degree of "quality" built into the resulting product.

Typically when a schedule is being formulated, the work to be done is broken down into tasks or work packages that are organized into a logical pattern usually called a *work breakdown structure* (WBS). The WBS is usually formulated in a hierarchical manner that may follow methodology established for a particular industry or organization. The amount and type of cost-to-complete each work packet is then estimated and resources to perform the work are identified, either generically or specifically. The estimated cost is typically a function of the amount and type of resources; dependent tasks are identified (a list of tasks that must be completed before starting this task).

These tasks are then typically input to a scheduling program that produces a time phasing of task start and end dates based upon the project start date, task resource needs, resource availabilities, and task interdependencies. When these tasks are rolled-up the WBS hierarchy, the total cost plan is derived, as shown in Figure 4.3. The U.S. government's term for this planned cost curve is "budgeted cost of work scheduled" (BCWS).

As the project progresses, actual cost are incurred by the effort expended in each work package and the total actual cost can be plotted as shown in Figure 4.4. Also, the relative amount of the things needed to be accomplished within the work packet that have actually been completed (% complete) can be determined or at least estimated. For example, if an activity had an estimated total cost of $10,000, and if the things to be done in the activity were 70 percent complete (or 70 percent of the elements were complete), then the earned value would be $7,000. Since percent complete and earned value can be estimated for each work packet, the total project earned value at a point in time can be determined by a WBS roll-up of the values.

The earned value is a point on the planned cost (BCWS) curve. This is illustrated in Figure 4.5, which shows the planned cost and actual cost curves for a project analysis. Variances between the three values BCWS (planned cost), BCWP (earned value), and actual cost (ACWP) yield the earned value metrics. There are earned value metrics available for both cost and schedule variances. The cost metrics are as follows:

- Cost variance (dollars) = BCWP − ACWP
- Cost variance (percent) = (BCWP − ACWP) × 100%/BCWP

FIGURE 4.3. THE PROJECT COST CURVE.

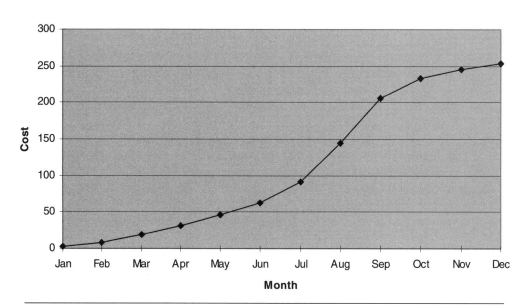

Project Cost Plan

FIGURE 4.4. COMPARISON OF ACTUAL (TRIANGLE) VERSUS PLANNED COSTS (DIAMOND).

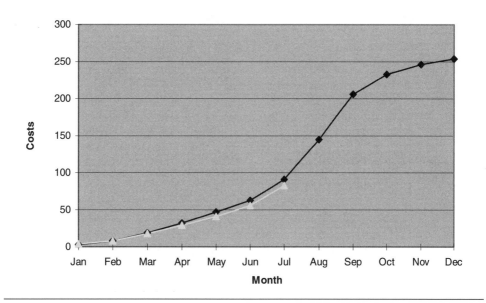

Plan and Actual Cost

- Cost efficiency factor or Cost Performance Index (CPI) = BCWP/ACWP
- Estimated cost at completion (EAC) = Budget at completion/(CPI)

There are several other EAC formulas, and the most appropriate depends upon project type and when the EAC is calculated (Christensen, 1995). The schedule metrics are as follows:

- Schedule variance (dollars) = BCWP − BCWS
- Schedule variance (months) = (BCWP − BCWS)/(Planned cost for month)
- Schedule efficiency factor or Schedule Performance Index (SPI) = BCWP/BCWS
- Estimated time to complete = (Planned completion in months)/(SPI)

The schedule variance in time (S) is shown in Figure 4.5 along the time axis.

Usually when project progress is reported, two types of information are presented: schedule data and cost data. Schedule data is typically shown in a Gantt or similar type chart, as shown in Figure 4.6 for an example project. Cost data is typically reported as actual costs versus planned costs at some upper level of the WBS. The cost variance is often just reported at the total level as total actual cost incurred versus budget.

FIGURE 4.5. GRAPHICAL DEPICTION OF EARNED VALUE.

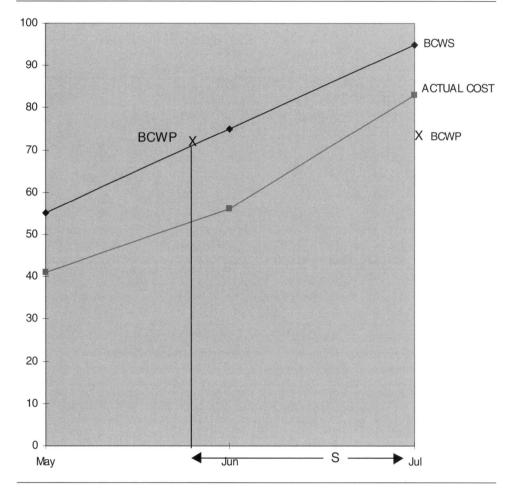

The problem with these usual methods is that they do not provide a clear quantitative picture of the true project status, nor do they provide a means for extrapolating project cost to complete or completion date. For example, when we look at the Gantt chart (which also shows the task % complete as dark bars stripes inside the bars) and we then see that we are not over budget (say, an actual cost of 83,000 versus planned cost of 91,000), it is hard to determine how much we are behind schedule and it *appears we are not overspending.*

However, on this project *we are well behind schedule and are overspending.* An earned value analysis would show that the schedule variance is 0.67 months (behind schedule) and the cost variance is $10,000 (overspent). The estimated time to complete is 15 months instead of the 12 months planned, and the estimated cost to complete is 289,000 instead of 254,000.

FIGURE 4.6. A GANTT CHART SHOWING SCHEDULE DATA.

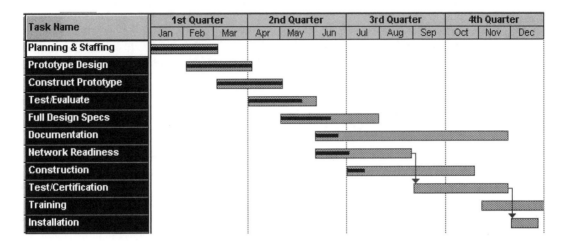

Task Name	1st Quarter			2nd Quarter			3rd Quarter			4th Quarter		
	Jan	Feb	Mar	Apr	May	Jun	Jul	Aug	Sep	Oct	Nov	Dec
Planning & Staffing												
Prototype Design												
Construct Prototype												
Test/Evaluate												
Full Design Specs												
Documentation												
Network Readiness												
Construction												
Test/Certification												
Training												
Installation												

The exact calculation of these numbers and their illustration in a spreadsheet is shown in Brandon (1998).

Often, *effective* implementation of earned value is difficult in some organizations. A common key problem involves data acquisition of "percent complete" data. Data on task completion must be obtained at regular intervals, but the effort involved in gathering the information must not be burdensome. It is best to set up this data acquisition as a by-product of some other required reporting mechanism such as weekly time card reporting.

Another common problem concerns the accuracy of the estimates versus the time required to calculate the estimate. As long as the project WBS is developed to point where each task at the lowest level of the WBS is only a person-week or so effort, then approximation techniques are sufficiently accurate and quick. There are a number of approximation techniques described in the EVA literature: One of the simplest uses 0 percent for a task that has not begun, 50 percent for a task that has started, and 100 percent for a task that is complete.

Another potential EVA problem area concerns employee resistance and the honesty of completion percentage reporting. EVA should not be used directly for employee evaluations; doing so will certainly compromise the main purpose of the system: project performance measurement. Computer systems to do EVA analysis are available, and an organization can always program their own (or set it up with spreadsheets). For example, a computer EVA system can look for employee reporting problems such as showing the same % complete on a task from week to week where hours have been charged to that task. Note that most off-the-shelf project scheduling software will not offer these EVA management analyses. These issues are also addressed in Brandon (1998).

Earned value methods have another advantage over current reporting techniques (Gantt charts and cost versus budget). Since earned values are quantitative numbers expressed in dollars or person-hours (for both cost and schedule deviations), these numbers can be rolled up, along an OBS for example, to give a picture of how all projects of varying sizes are performing in an organization. Since underspent projects do not necessarily help overspent projects (in either time or dollars), often the positive variations are set to zero and estimate at completion (EAC) is unchanged. This is illustrated in Figure 4.7. If spreadsheet models are used for earned value, then these are easily interfaced with most executive information systems.

Success Criteria and Success Factors

EVA methodology is based upon time and cost metrics for a given scope and quality. Cost, time, and quality (often referred to as the "iron triangle") have formed the prime basis for measuring project success for the last 50 years (Atkinson, 1999). However, a number of authors in more recent years (Morris and Hough, 1987; Pinto and Slevin, 1988; DeLone and McLean, 1992; Lim and Mohamed, 1999; and Atkinson, 1999) have suggested that other criteria for success are also important. Some of these other criteria may be less quantitative and more difficult to measure. Also some of these criteria may be temporary in that their values may be much more important at some points in the project, but less important at other points like the end of the project.

Many of these authors raise the question of "what is a successful project." Different stakeholders involved with the same project may have different opinions about a project's success. One example given concerns the construction of a shopping center that is eventually

FIGURE 4.7. AN EARNED VALUE SPREADSHEET FOR MULTIPLE PROJECTS.

PROJECT	BCWS	BCWP	- TIME VAR ($)	VAR +	ACWP	- COST VAR ($)	VAR +	PLAN	EAC
Project 1	91	73	18	18	83	10	10	254	289
Project 2	130	135	-5	0	125	-10	0	302	302
Project 3	65	60	5	5	75	15	15	127	159
Project 4	25	23	2	2	27	4	4	48	56
Project 5	84	82	2	2	81	-1	0	180	180
Project 6	53	47	6	6	48	1	1	110	112
Project 7	102	103	-1	0	110	7	7	190	203
Project 8	35	37	-2	0	40	3	3	78	84
	585	560			589				1385

Total Schedule Overage	33	Total Cost Overage	40
Relative Schedule Overage	5.64%	Relative Cost Overage	6.84%
Schedule Overage (Months)	0.68	Cost at Completions	1385

completed to the quality standard—however, with significant cost and time overruns. Some stakeholders are very unhappy, and that depends upon the type of contracts involved and which party(s) contractually bear the burden of the cost overruns. Other stakeholders, such as the public using the mall and the merchants renting space in the mall and the government getting tax revenue from the mall, are pleased with the results and see the project as a great success. They define two perspectives: the macro perspective, which involves all the stakeholders, and the micro perspective, which involves only the construction parties, such as the developer and contractors. The macro perspective is relevant for all phases of a project from conceptualization through construction and then operation. The micro perspective is most relevant for the construction phase. This is a theme explored extensively later in this book.

Lim and Mohamed (1999) define two types of success criteria: completion and satisfaction. Completion criteria include contract-related items such as cost, time, scope, and quality. Satisfaction criteria include utility (fitness for purpose) and operation (ease of use, ease of learning, ease of maintenance, etc.). The macro perspective involves both completion and satisfaction criteria; the micro perspective only involves completion perspectives. This is illustrated in Figure 4.8.

Lim and Mohamed draw a clear distinction between "success criteria" and "success factors." The criteria are "a principle or standard by which anything is or can be judged"; factors are "any circumstance, fact, or influence which contributes to a result". Figure 4.9 from their article illustrates this point. Factors for the completion criteria would typically include resource variables (cost, availability, skill, motivation, etc.), management variables (project manager skill, line management support, etc.), and risk variables (weather, economy, technology, etc.). Factors for the satisfaction criteria would be those things that drive the satisfaction of the stakeholders.

Success criteria tend to be relatively independent of the type of project being measured. However, the factors are very dependent on the type of thing being built (or accomplished). For example, a factor for the satisfaction type criteria of utility (using the shopping center model) might be "ample parking"; a factor for the operation factor of the satisfaction criteria might be "ease of parking".

DeLone and McLean (1992) developed a taxonomy of success criteria and factors for information systems. The six major criteria are system quality, information quality, use

FIGURE 4.8. A GENERALIZATION OF LIM AND MOHAMED'S TWO TYPES OF SUCCESS CRITERIA.

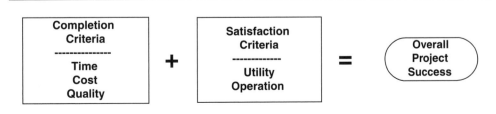

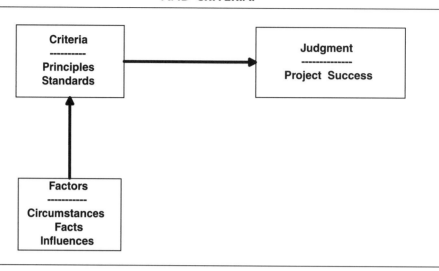

FIGURE 4.9. LIM AND MOHAMED'S RELATIONSHIP BETWEEN FACTORS AND CRITERIA.

(amount), user satisfaction, individual impact, and organizational impact. Figure 4.10 illustrates this taxonomy: system and information quality foster use and satisfaction, which determine individual impact, which then determines organizational impact. They did not divide these into completion and satisfaction types, but the first two would likely be more completion-oriented, and the last four more satisfaction-oriented. Their taxonomy listed a number of factors (which they called measures) for each criterion. Molla and Licker (2001) recently extended DeLone and McLean's taxonomy to e-commerce systems. Their success criteria are system quality, content quality, use, trust, and support.

In terms of project performance measurement, both success criteria and factors are important and useful. Even the less quantitative factors can be used as an early-warning system to potential problem areas. We need to first determine our concern and perspective: macro or micro. Then we need to list our success criteria. Next the success factors for each criterion need to be determined. Then we need to set up feedback systems using our understanding of the cause-and-effect relationship between the criteria/factors and our management decision rights over adjustable items (resource type, compensation, etc.). For completion criteria (time, cost, quality, scope, performance, safety), EVA is an excellent performance tool. For the less quantitative factors (satisfaction, utility, etc.), some other metrics have to be employed. Many of these less quantitative success criteria and factors are also considered in PMI's PMBOK. PMBOK processes are defined to manage these factors, particularly in the area of communications, human resources, and risk. Metrics for these factors may involve formal or informal surveys, and these are discussed later in the *Balanced Scorecard (BSC) Approach* section coming up. Consideration of both the completion

FIGURE 4.10. DELONE AND MCLEAN'S TAXONOMY OF SUCCESS CRITERIA AND FACTORS.

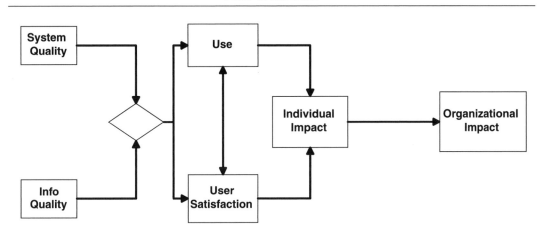

criteria and the satisfaction criteria leads to a more balanced approach toward project success and performance measurement.

Balancing Time/Cost with Satisfaction

EVA methods offer key management information to take corrective action primarily at the short-term tactical level for a particular project. When EVA metrics from all projects in a program are combined, management has a greater visibility to examine issues across multiple projects and take actions—some of which may be strategic as well as tactical. Project management is often criticized as being too shortsighted and limited to tactics that may not be in the best interest of the long-term success of an organization. Certainly, we understand (and most of us have seen cases) where employees are pushed too hard for success of one project only to become demoralized for performance on future projects.

Organizations should consider implementing a "corporate project management strategy". Implementing a strategy involves the development and/or adoption of some type of strategic foundation upon which to translate corporate vision and goals into winning strategies with effective tactics. Long-term winning strategies must address both the quality and satisfaction dimensions, as well as the financial and temporal dimensions.

A number of strategic foundations have been used by organizations in the past, but lately the balanced scorecard approach seems most successful and relevant in today's global economic setting. We know that today's economic setting is often typified by the phrase "better-cheaper-faster." We also realized that companies compete is a global economy today, and this is becoming even more so with e-commerce and the very rapid progression in technology.

However, the two main factors in the increased need for consideration of the quality and satisfaction dimensions in overall performance are the fact that traditional project management techniques are mostly tactical, not strategic, and the fact that the corporate value of intangible versus tangible assets has been shifting considerably in the last 20 years. Note that in disciplines that have always had problems with project performance (such as information technology), there were always a high percentage of intangible assets.

Balanced Scorecard (BSC) Approach

The BSC is an approach to strategic management that was originally proposed in 1992 by Kaplan and Norton (Kaplan, 1992). They recognized some of the problems with classical management approaches including, traditionally, an overemphasis on metrics that were strictly financial based and that looked mostly at the results of management decisions in the past. This rear-facing approach was becoming obsolete in today's fast-moving economic and technology-based economy. The balanced scorecard approach provides definitive procedures as to what companies should measure in order to balance this traditional prime focus of solely a financial perspective.

While the processes of modern business have changed dramatically over the past several decades (particularly in the last five years with the growth of the Internet), the methods of performance measurement have stayed much the same. Past measurement methods were well suited to asset-based, slow-changing manufacturing organizations. But past performance measurement systems are no longer as relevant to capture the value-creating mechanisms of today's modern business organizations (Kaplan, 1996). Today intangible assets such as employee knowledge, customer base and relations, supplier base and relations, and access to innovation are the key to creating value; Figure 4.11 shows this shift for U.S. companies. When we view financial statements (profit/loss, balance sheet, cash flow), Gantt charts, or even EVA analysis, we are seeing the results of actions and decisions that occurred in the past ("lagging indicators") (Eickelmann, 2003). Some of these past actions and decisions that determined an organization's present financial state may have taken place a month ago, a year ago, or a decade ago.

The balanced scorecard does not ignore financial matters but changes the perspective from one of reaction to one of proactive involvement. The balanced scorecard approach takes a look at the key management actions (including metrics) that will most likely affect a company's financial state *in the future* ("leading indicators") (Eickelmann, 2003). It has organizations first clearly and quantitatively define their vision and strategy, and then turn them into measurable actions. It provides feedback around both the internal business processes and external outcomes to continuously improve overall performance. When

FIGURE 4.11. THE GROWTH IN INTANGIBLE ASSETS IN U.S. COMPANIES SINCE THE 1980S.

Intangible Assets

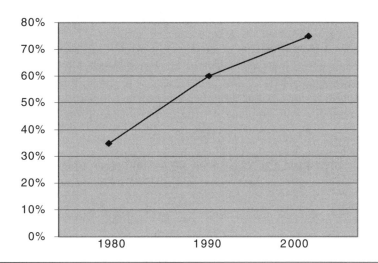

appropriately implemented, a "balanced scorecard approach transforms strategic planning from an academic exercise into the primary control mechanism" of an organization (Niven, 2002; Averson, 2002).

In the language of the founders of BSC (Kaplan and Norton):

> The balanced scorecard retains traditional financial measures. But financial measures tell the story of past events, an adequate story for industrial age companies for which investments in long-term capabilities and customer relationships were not critical for success. These financial measures are inadequate, however, for guiding and evaluating the journey that information age companies must make to create future value through investment in customers, suppliers, employees (Kaplan, 1996)."

Today BCS has been implemented successfully in thousands of organizations around the world, both "for-profit" and "nonprofit." Reengineering success ratios using BSC are reported as (Averson, 2002):

- Nonmeasurement-managed organizations (55 percent)
- Measurement-managed organizations (97 percent)

The BSC approach defines four "perspectives" from which an organization is viewed. Metrics are developed, then data collected and analyzed relative for each of these perspectives. This philosophy is illustrated in Figure 4.12.

The first perspective is Learning and Growth. This involves the investment in human capital through activities like positive feedback, motivational techniques, setting up mentors, communications facilitation, employee training, and development of a "company culture" supporting quality. In today's knowledge worker organization, employees (the main repository of know-how) are the main resource. In the current climate of rapid technological change, it is becoming necessary for knowledge workers to always be learning—lifelong education. Organizations often find themselves unable to hire new technical people and at the same time have reduced training of existing employees. This is a leading indicator of brain drain that can ultimately destroy a company (Averson, 2002). Metrics can be put into

FIGURE 4.12. THE FOUR MEASUREMENT PERSPECTIVES OF THE BSC APPROACH.

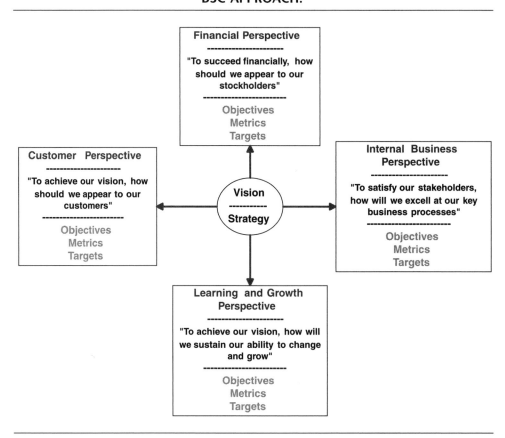

place to guide managers in focusing personnel development funds where they can help the most. This perspective is the foundation for long-term success of an organization.

The next perspective is the Business (or Internal) Process. This relates to the internal methods used to produce the goods and/or deliver the services. Metrics based on this perspective provide managers with knowledge on how well their business is running and is fine-tuned, and whether the products and services conform to customer-stated requirements and unstated expectations. These metrics have to be carefully designed by those who know these processes best. There are two kinds of business processes in an organization's value chain: primary (mission-oriented) processes, which are the unique functions of an organization, and support processes (accounting, legal, procurement, HR, etc.), which are more repetitive and common and hence easier to measure and benchmark using classical metrics.

The next perspective is the Customer. Recent management philosophy and the popularity of IT products as CRM (customer relations management systems) and SFA (sales force automation systems) have demonstrated an increasing appreciation of the importance of customer focus and customer satisfaction in any business. If customers are not satisfied, they will eventually find other companies that will meet their needs. Poor performance from this perspective is also a leading indicator of future decline, even though the current financial picture may look rosy. From a project management perspective, some customer relations are included in the PMI process groups and knowledge areas, particularly for external projects.

The last perspective is Financial. Kaplan and Norton do not ignore the traditional importance of financial data. Timely and accurate accounting data will always be important. Their point is that the current emphasis on financials leads to the unbalanced situation with regard to other perspectives. Today there is also a need to include additional financial-related data, such as risk assessment and cost-benefit data, in this category. Again from a project management perspective, financial data is a key metric, including EVA.

Figure 4.13 shows the perspectives in a waterfall, long-term, strategic cause-and-effect scenario. Motivation, skills, and satisfaction of employees are the foundation for all improvements. Motivated, skilled, and empowered employees will improve the ways they work and also improve the work processes. Improved work processes will lead to improved products and services that will mean increased customer satisfaction. Increased customer satisfaction will lead to long-term improved financial performance.

(The BSC methodology incorporates some key notions of Total Quality Management (TQM) such as customer-defined quality, continuous improvement, employee empowerment, and measurement-based management and feedback. In the early industrial revolution, quality control and zero defects were big management buzzwords. To prevent the customer from receiving poor-quality products and services, stringent methods and efforts were put into inspection and testing at the end of the production line. The problem with this approach, as pointed out by Deming (Walton, 1991), is that the true causes of defects can never be identified, and there will always be inefficiencies because of the rejection of defects. What Deming saw was that variation is created at each point in a production line, and the causes of defects need to be identified and corrected. Deming emphasized that all business

FIGURE 4.13. THE WATERFALL RELATION BETWEEN THE BCS MEASURES.

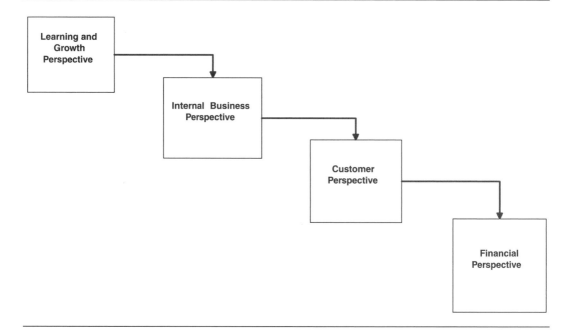

processes should be part of a system with feedback loops. The feedback data should be examined by managers to determine the causes of defects (the processes with significant problems), and then attention can be focused on fixing those particular problem processes. BSC emphasizes feedback control from each of the four perspectives in each major business process; that is, each perspective for each key process should have objectives, metrics, and targets for proper feedback control.)

Integrating EVA and Success Factors into BSC to Maximize Performance

How would one integrate the successful concepts of BSC in a project management context (Phillips et al., 2002)? Considering a project as a key business project, one needs to establish metrics for each of the four perspectives at the individual project level and also at the PMO (project management office) level. (See the chapter by Young and Powell for a discussion of the PMO.) EVA would be a key tool for the BSC financial perspective. Success factors from the "completion" perspective could be used in the BSC internal perspectives. Success factors from the "satisfaction" perspective could be used in the BSC customer perspective. This is illustrated in Figure 4.14.

FIGURE 4.14. THE POSITION OF EVA AND SUCCESS FACTORS IN THE BALANCED SCORECARD.

```
+---------------------+
|   Learning  and     |
|      Growth         |
|   ------------       |
|                     |
+---------------------+


+---------------------+
|     Customer        |
|   -----------       |
|    Satisfaction     |
|      Factors        |
+---------------------+


+---------------------+
|     Internal        |
|   -----------       |
|    Completion       |
|      Factors        |
+---------------------+


+---------------------+
|     Financial       |
|   -----------       |
|       EVA           |
+---------------------+
```

A scorecard defines the key activities for each perspective and the metrics that are going to be used to monitor the performance of that activity. One possible scorecard is shown in Figure 4.15 for the project level and for the PMO level. For example, at the BSC customer perspective, we would measure not only the customer's satisfaction with the system but also how involved our project team was with customer personnel. These metrics could be determined both at project completion and also at project milestones for early warning of potential problems.

The integration of BSC concepts into PM could also eventually lead to extensions of the PMI process groups and knowledge areas of Figure 4.1—that is, PMBOK.

Summary

EVA, critical success factors, and BSC are some of the most powerful and successful management tools for measuring and increasing performance. However, these techniques are so far rarely if ever used together, and proponents from one camp often criticize the methods of the other, probably because of a lack of full understanding of the other's approach.

FIGURE 4.15. BALANCED SCORECARD EXAMPLE.

Project Management BSC Metrics.

Perspective	Activities	Metrics	PMI Knowledge Area
Learning and Growth	Effective Communications Team Building and Motivation Employee Development Mentor Program	Communication Problems Employee Feedback Training Funding Mentor Reviews	Communications Human Resources Human Resources Human Resources
Customer	Customer Relations Customer Involvement	Customer Satisfaction Customer Interaction Logs	Communications Scope
Internal	Planning Sizing "Built-in Quality" Testing and Inspection Scope Control	EVA Time, SPI Estimation Accuracy Standards Methodology Specifications, QA Change Orders	Time and Risk Time Scope Risk Quality Scope
Financial	Use of Resources Contracting/Procurement	EVA Cost, CPI Procurement Standards	Cost Procurement

PMO BSC Metrics.

Perspective	Activities	Metrics
Learning and Growth	Effective Communications Management Team Building and Motivation Project Manager Development Motivation PM Forum	Communication Problems PM and Employee Feedback Training Funding PM Incentive Packages Sharing of "Lessons Learned"
Customer	Customer Relations Customer Involvement	Customer Satisfaction Customer Feedback (about PM)
Internal	Planning PM Support (from line management)	Roll-up of EVA Time PM Feedback
Financial	Use of Resources Contracting/Procurement	Roll-up of EVA Cost Procurement Problems

EVA, critical success factors, and BSC need not be opposing management techniques. Tactical techniques like EVA and critical success factors can be used as some of the key metrics for the strategic BSC approach.

This chapter has presented one way that the approaches can be fully integrated into the managing of projects.

References

Atkinson, R. 1999. Project management: Cost, time, and quality. *International Journal of Project Management* 17(6):337–342

Averson, P. 2002. The Balanced Scorecard Institute. www.balancedscorecard.org.

Brandon, D. 1998. Implementing earned value easily and effectively. *Project Management Journal* 29(2).

———. 1999. Implementing earned value easily and effectively. In *Essentials of project control*, 113 ff. Newtown Square, PA: Project Management Institute.

Cale, E. G., J. R. Curley, and K. F. Curley. 1987. Measuring implementation outcome: Beyond success and failure. *Information and Management* 3(1):245–253.

Cafasso, R. 1994. Few IS projects come in on time, on budget. *Computerworld* 28(50):20.

Christensen, D., and D. Ferens. 1995. Using earned value for performance measurement on software projects. *Acquisition Review Quarterly* (Spring): 155–171

Christensen, D. S., et al. 1995. A review of estimate at completion research. *Journal of Cost Analysis* (Spring).

DeLone, W., E. McLean. 1992. Information systems success: The quest for the dependent variable. *Information Systems Research* 3(1):6–95.

Eickelmann, N. 2003. Achieving organizational IT goals through Integrating the balanced scorecard and software measurement frameworks. Chapter II in *Technologies and Methodologies for Evaluating Information Technology in Business*, ed. C. K. Davis. Hershey, PA: IRM Press.

Field, T. 1997. When bad things happen to good projects. *CIO* 11(2):54–62

Fleming, Q., and J. K. Koppelman. 1994. The essence of evolution of earned value. *Cost Engineering* 36(11):21–27

———. 1996. *Earned value project management.* Newtown Square, PA: Project Management Institute.

———. 1998. Earned value project management: A powerful tool for software projects. *CROSSTALK* (July).

Hewitt, L., and M. O'Connor. 1993. Applying earned value to government in-house activities. *Army Research, Development & Acquisition Bulletin.* (January–February): 8–10.

Hildebrand, C. If at first you don't succeed. *CIO Enterprise.* Section 2:4–15.

Horan, R., and D. McNichols. 1990. Project management for large systems. *Business Communications Review* 20 (September): 15–24.

Kaplan, R., and D. Norton. 1996. *The balanced scorecard.* Cambridge, MA: Harvard Business School Press.

———. 1992. The balanced scorecard: Measures that drive performance. *Harvard Business Review* (January).

———. 1993. Putting the balanced scorecard to work. *Harvard Business Review* (September).

Kiewel, B. 1998. Measuring progress in software development. *PM Network* (January): 29–32.

Lewis, J. 1995. *Project planning, scheduling, and control.* Chap. 10. Homewood, IL: Irwin.

Lim, C. S., and Z. Mohamed. 1999. Criteria of project success: An exploratory re-examination. *International Journal of Project Management* 17(4):243–248.

Molla, A., and P. Licker. 2001. E-commerce systems success: An attempt to extend and respect the DeLone and Maclean model of IS success. *Journal of Electronic Commerce Research.* 2(4):131–141.

Morris, P. W. G., and G. H. Hough. 1987. *The anatomy of major projects.* Chichester: Wiley.

Niven, P. 2002. *Balanced scorecard: Step by Step.* New York: Wiley.

Phillips, J. J., T. W. Bothell, and L. Snead. 2002. *The project management scorecard: Measuring the success of project management solutions.* Oxford, UK: Butterworth-Heinemann.

Pinto, J., and D. Slevin. 1998. Critical success factors across the project lifecycle. *Project Management Journal* XIX:67–75.

U.S. Department of Commerce. 1999. Guide to balanced scorecard performance management methodology (July). Available at http://oamweb.osec.doc.gov/bsc/guide.htm.

Walton, M., and W. E. Deming. 1991. *Deming management at work.* New York: Perigee, 1991.

Yeates, D., and J. Cadle. 1996. *Project management for information systems,* London: Pitman.

CHAPTER FIVE

QUALITATIVE AND QUANTITATIVE RISK MANAGEMENT

Stephen J. Simister

Risk is present in all projects, and project managers are routinely involved in making decisions that have a major impact on risk. Risk management is concerned with establishing a set of processes and practices by which risk is managed, rather than being dealt with by default. The effective management of risk can only be achieved by the actions of the whole project team, including the client.

Risk management formalizes the intuitive approach to risk that project teams often undertake. By utilizing a formal approach, project teams can manage risk in a more proactive manner. In addition, there needs to be an overall risk management strategy so that this risk management process is implemented in a coordinated fashion. This strategy should include how risk management will be integrated into the project management process on a project.

Definitions

In the context of this chapter, risk management is considered to be a process for identifying, assessing, and responding to risks associated with delivering an objective—for example, completing a construction project—and the focus is on commercial-type risks. Health- and safety-related risks are likely to need separate consideration; Ward and Chapman look at this in their chapter later in the book.

The risk management terms used in this section follow the International Organization for Standardization (ISO) and ISO and International Electrotechnical Commission (IEC)

Guide to Risk Management Terminology (ISO, 2002). The guide covers 29 terms and definitions for risk management, which are categorized into one of four groups: basic terms, terms related to people or organizations affected by risk, terms related to risk assessment, and terms related to risk treatment and control.

The Rationale for Risk Management

Project managers will invariably be called upon to advise as to whether they should undertake risk management on their project. While the benefits of risk management may not be immediately apparent, the direct costs of undertaking the process will be.

The demands of delivering a project are extremely onerous. The desire to deliver projects cheaper and faster while moving closer to the boundaries of innovative design increases the risk profile of a project. In this respect, risk is good: without it there would be no opportunity. It is the very presence of risk that represents the opportunity for a project to go ahead in the first place. Furthermore, the actual impact of risk is unique to a project even though its presence may be commonplace.

Risk management should be flexible, adapting to the circumstances of the client's needs and the project. Some clients require a snapshot of the risks at the outset of the project, with an initial risk assessment, the provision of a one-off risk register, and a quick estimate of the combined effect. Other projects may require a full risk management service, with risk being continually addressed throughout the project.

The Recognition of Uncertainty

By undertaking risk management, the project manager can ensure that clients appreciate just how sensitive their projects are to changing circumstances. This leads to the following:

- Increased confidence in achieving the project objectives and therefore improved chances of success.
- Surprises being reduced, such as cost or time overruns or forced compromises of performance objectives. These surprises usually result in "fire fighting" and the ineffective application of urgent remedial measures.
- Identification of opportunities as risks are diminished, perhaps by relaxing overcautious practices such as duplicating insurances or seeking performance bonds unnecessarily.
- All ranges of parameters that might affect the project being incorporated, rather than a single-point estimate (for example, the cost for a particular component can be between $200 and $300, rather than $225).
- Allowing the team to recognize and understand the composition of contingencies, thus avoiding duplicating any allowance already made.

Justifiable Decision Making

Risk management allows decision making to be based on an assessment of known variables. Judgment can become far more objective and justification for action (or inaction) demonstrable both at the time and at a later date, should it be questioned. As part of risk management, risk is often transferred from one party to another—for example, the main contractor to the subcontractor. The formal application of risk management should ensure that risk transfer is based on the rational assessment of a party's ability to bear and control the risk.

There is a wide range of techniques that can be utilized to facilitate effective decision making. The use of risk management provides an audit trail that can be used should problems occur during later stages of the project.

Team Development

During risk management workshops, a "snapshot" of what might happen to the delivery of a project is taken. At the workshops the project team discuss their concerns and agree on a common way forward. The team must also explore the consequences of risks occurring. Such consequences can be discussed without fear of penalty or contractual restrictions. Discussing the consequence of risk in this way normally reinforces to the team that the project objectives can only be achieved if they work as a team, and not as self-interested parties. As with other types of facilitated workshops, such as value management, such team development-related benefits could on their own justify undertaking risk management.

In summary, to fully realize the benefits of risk management, certain principles should be adhered to:

- Risk management is an awareness of uncertainty.
- Awareness of uncertainty can lead to positive outcomes if recognized early enough.
- Risk management will add both direct and secondary benefits to the delivery of any objective.
- The management of risk must not remove incentive.
- Allocation of risk is to be gauged by the various parties' abilities to bear and control that risk.
- Complete transfer of risk is rarely wholly effective or indeed possible.
- Information about and perception of a risk are fundamental to its assessment and acceptance.
- All risks change with time and any action (or inaction) taken (or not taken) upon them.

The Risk Management Process

A number of risk management processes are publicly available. These can be broken down into three groups: those issued by national standards associations, those issued by professional institutions and those issued by government departments. Information on the various publications is provided in the further reading section at the end of this chapter. In addition, the Web site address for the organizations is also provided.

National standards associations

- *British Standards Institute (2000)*. The UK national standards association. The risk management standard forms part of a wider project management standard (www.bsi-global.com).
- *Canada Standards Association (1997)*. Canada's national standards association (www.csa.ca).
- *Standards Australia (1999)*. Australia's national standards association (www.standards.com.au).

Professional institutions

- *ICE (1998)*. Produced by the UK-based Institution of Civil Engineers (ICE) in partnership with the Institute of Actuaries. The risk management process is designed specifically for infrastructure projects, such as roads and so on (www.ice.org.uk).
- *Japan Project Management Forum (2002)*. This Japan-based forum describes risk management as an input into how project management can best stimulate innovation and generate improved business value to a company (www.enaa.or.jp/JPMF/).
- *PMI (2000)*. The Project Management Institute's *A Guide to the Project Management Body of Knowledge* has a chapter on risk management (www.pmi.org).
- *APM (Dixon, 2000)*. The Association for Project Management's body of knowledge includes a section on risk management. The APM also produces a specific risk management guide (Simon, Hillson, and Newland, 1997; www.apm.org.uk/).

Government departments

- *DoD (2002)*. The U.S. Department of Defense has a process showing specifically how risk management is to be applied to defense projects (www.defenselink.mil/).
- *The UK Office of Government Commerce (2002)*. The OGC has a generic guide to managing risk in a project environment (http://www.ogc.gov.uk)

All these publications propound their own risk management process. While the processes differ from each other to some extent, there is a common process that runs through them all. This common risk management process consists of five basic steps: risk strategy, risk identification, risk analysis, risk response, and risk control. These five steps are iterative in nature, and it is this iteration that constitutes risk management.

The five steps are detailed in the following:

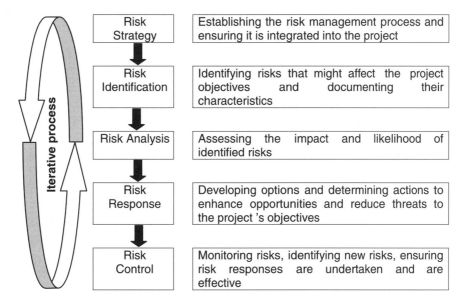

The process proposes a formal application of risk management. The dangers of an informal process are that stages are missed. Projects are typically undertaken within demanding time frames. In such circumstances there is a temptation to deal with risk reactively. In such cases it is only possible to deal with the consequences of the risk occurring; there is no opportunity to mitigate or even avoid the risk.

The risk management process should be commenced as early as possible in a project life cycle. Since any risk management assessment is a snapshot in time, the process has to be undertaken on an iterative basis.

Risk Strategy

The risk strategy needs to set out how risk management will be undertaken on a project. The risk strategy needs to be integrated with the project strategy as well as the wider project management process.

As part of the risk strategy, a risk management plan should be developed. The risk management plan is similar in nature to the project execution plan and may form part of that document. Key areas of a risk management plan might include the following (Simon, Hillson, and Newland, 1997):

- Scope and objectives of risk process
- Roles and responsibilities of participants in the process
- Approach & process to be used
- Deliverables of the process
- Review and reporting cycle
- Tools to be used

The risk strategy should also describe how the risk management process would contribute to any project evaluation exercise that might take place at the completion of the project.

Risk Identification

The identification stage should commence as early as possible, preferably as part of any feasibility study for the project. One of the key principles is that the identification of risks should be undertaken at various stages of the project life cycle. Identification is not just undertaken at the start of the project and then simply monitored during execution. The execution phase itself will generate new risks that need to be identified through the risk identification process.

Risk Identification Methodology

(*i*) *Structured identification of all sources of risk to the project.* The risk management process is based on the concept of providing a structured approach to dealing with risk. It may seem obvious that the availability of suitably qualified labor is one source of risk, but how would this impact other risks that might be identified for the project? It is this interrelationship between risks that can only be drawn out through the use of a structured workshop and identification methods, where the whole project team has the ability to contribute.

(*ii*) *Preliminary analysis to establish probable major risks for further investigation.* The identification stage is designed to prompt team thinking and to bring out all possible risks that might impact the project. Several hundred risks could be identified, and the team may not have time to analyze and develop responses to them all. It is normal during the identification stage to begin by sifting and removing those risks that the team feels are not worth investigating further.

(*iii*) *The true risk needs to be identified.* Risks are caused by background conditions that are essentially "givens," for example, location of site. It is these background conditions that cause risks to occur. If the risk does occur, it will have an effect on the project. This is shown diagrammatically in Figure 5.1.

It is all too easy to mistakenly manage the effect rather than the risk itself. When the team identifies and manages the true risk, their effort is expended managing the risk rather than the effect.

Risk Identification Techniques

(*i*) *Research.* While exactly the same project will not have been executed before, something similar will have been. Investigation into projects on neighboring sites or projects previously undertaken by the client should be instigated.

(*ii*) *Structured interviews/questionnaires.* Interviews with key members of the project team (including client and suppliers) will elicit the greatest insight into risks to the project and how they are perceived by individuals. Interviews allow a greater depth of understanding to be achieved than group discussions but do take up a lot of time.

FIGURE 5.1. CAUSE, RISK, AND EFFECT.

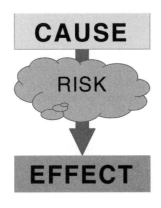

	Question	Example
CAUSE	What background condition is driving the risk?	The project is being undertaken in a particular city.
RISK	What is the area of uncertainty?	There may not be enough suitably qualified labor in the city.
EFFECT	What is the consequence of the risk occurring?	Additional labor from outside the city is required, which adds cost to the project.

(iii) Checklists/prompt lists. A simple and effective way to stimulate the team into thinking about risk is to use a checklist. An easy way to start a checklist is for each team member to write down all the variation orders on their last project, with the reasons for their issue.

(iv) Brainstorming in a workshop environment. Bringing the team together in a focused workshop is a powerful environment in which to discuss risk. The team can have a better understanding of how each member perceives risk differently. One important aspect of such workshops is that lateral thinking is encouraged. The workshop environment gives the team an opportunity to experiment with different viewpoints that individuals might normally reject out of hand if working alone.

(v) Risk register. A risk register is probably the most useful tool in the risk management process. It enables risks to be logged and tracked through the life of a project. While a risk register can be maintained by hand, it is much more useful to use a computer spreadsheet package. This allows the information to be sorted into different categories. Specialist risk management software is also available that integrates the risk register with risk analysis tools.

Typical column headings for a risk register are as follows:

- *Risk number.* A unique identifying number for the risk.
- *Risk description.* A written description of the risk.
- *Ownership.* Who is responsible for the management action in responding to the risk?
- *Probability.* How likely is the risk to occur?
- *Impact.* What happens if the risk does occur?
- *Risk factor.* Probability multiplied by impact.
- *Response.* What actions need to be taken to deal with the risk?

- *Status.* The status of the risk can be shown as:
  - *Done.* The risk has arisen and been dealt with.
  - *Active.* The risk is currently being managed.
  - *Monitor.* The risk has been identified, but no analysis or response has yet been developed for it.
- *Comments.* Allows notes to be kept on the risk.

It is also possible to place identified risks into categories, for example, "client-retained risks." The use of categories allows risk to be bundled together and can help responses to be tailored to deal with a category, rather than an individual risk.

An example risk register for a construction project is shown in Figure 5.2.

The risk register is a very useful format for showing a wide range of information in the risk management process. If the register is placed on a computer spreadsheet, it is easy to sort the risks in a variety of ways—for instance, by category or magnitude of probability. The risk register allows risks to be logged but can also display additional information linked to the project management process.

(vi) Database of historic risks. Experience is a great teacher, and the provision of a risk register allows information to be stored in a convenient format ready for use on the next project. The benefit of historical risk registers is that you know if a risk actually occurred and whether the appropriate response was set in place. The historic database should be used as a starting point in the risk management process, but although it may save some time in the identification stage, this stage cannot be completely omitted.

Risk Analysis

The analysis stage is concerned with trying to achieve a better understanding of the nature of the risks identified during the previous stage. After analysis, it will be possible to directly compare risks on a like-for-like basis. This is crucial in establishing the prioritization of risks in order to best apply organizational resources where they are most needed or can provide the biggest positive impact.

Risk analysis is generally divided into two parts: qualitative and quantitative. A qualitative risk analysis should always be undertaken as part of the risk management process. As part of the risk strategy, consideration should be given as if a quantitative analysis is required.

Qualitative Risk Analysis Methodology. Some form of qualitative risk analysis should be undertaken on all projects. It is the most basic form of risk analysis and is the foundation in understanding project risks.

Qualitative risk analysis can be performed at a number of levels. On the simplest level, the project manager can sit down with a few key stakeholders and discuss how risk should be managed on the project. A complex approach might involve having a dedicated subteam who manage the risk management process on behalf of the project manager.

No matter what level is chosen, a number of techniques can be used.

FIGURE 5.2. EXAMPLE RISK REGISTER.

ID	Risk Description	Ownership	Current Status D=Done A=Active M=Monitor	Probability Very high: 0.9 Very low: 0.1	Impact Very high: 0.8 Very low: 0.05	Risk Factor (Prob. x impact)	Response
100	Third-party influence						
110	Revisions to planning permission may be required.	Architect	D	0.7	0.05	0.035	Discussions with planning authority.
120	Contaminated ground hot spots may have to be removed from site.	Civil engineer	A	0.5	0.1	0.05	Discussions with environment agency.
130	Fire officer requirements not yet fixed.	Design team	A	0.7	0.1	0.07	Discussions with fire officer.
200	Client influence						
205	Protracted decision process — Client may not agree designs in line with design program.	Project manager	M				No response yet in place.
300	Design team						
310	Consultant appointments may not provide a clear definition of responsibilities.	Project team	D	0.1	0.05	0.005	Review consultants agreements.
320	Structural unknowns: composition of existing building (walls, roof).	Design team	M				No response yet in place.
325	Existing electrical power feed into building may not be of sufficient capacity.	Design team	M				No response yet in place.

(i) *Analysis of risks to assess the severity of impact and the probability of occurrence.* Risks have two dimensions that need to be assessed and analyzed:

- *Probability.* This is the likelihood of the risk occurring and is generally expressed as a percentage.
- *Impact.* If the risk did occur, what impact would it have on meeting the project's objectives?

For each of the risks identified, its probability and impact need to be assessed, normally in a workshop environment with key stakeholders present. In addition, before the risks can be entered into the risk register, a scale has to be set for the probability and impact dimensions. Setting a common scale will ensure a consistent approach to placing newly identified risks on the matrix at a later stage in the project life cycle. An indicative layout for defining scales is shown in Figure 5.3.

While the probability scale shown in Figure 5.3 may be utilized for any project, the impact scale needs to be set for each project. In Figure 5.3, the cost and time impacts are shown as a percentage of the original cost and time. Hence, a very high impact would mean the cost of the project would increase by more than 50 percent of the original budget.

Although time and cost impacts have been shown separately, it is possible to combine them, or to use other criteria such as performance. Again, the scale has to be set according to the needs of a project. Once set, the scales should not be changed during the life of a project. This allows risks to be compared on a common scale throughout a project and for the risk profile of a project to be accurately monitored.

A higher level of sophistication can be applied to the ranking of the risks on the probability/impact matrix. When generic units are assigned to a matrix, the relative importance of each cell can be seen, as shown in Figure 5.4.

The method of having the probability scale linear and the impact scale nonlinear emphasizes the significance of low-probability/high-impact risks. The matrix allows each risk

FIGURE 5.3. SCORING OF PROBABILITY AND IMPACT SCALES.

		IMPACT	
	PROBABILITY	Cost	Time
Very high	70%–95%	>50%	>50%
High	50%–70%	20–50%	20–50%
Medium	30%–50%	10–20%	10–20%
Low	10%–30%	5–10%	5–10%
Very low	5%–10%	<5%	<5%

FIGURE 5.4. MATRIX OF PROBABILITY VERSUS IMPACT.

PROBABILITY							
	Very high	0.9	0.045	0.09	0.18	0.36	0.72
	High	0.7	0.035	0.07	0.14	0.28	0.56
	Medium	0.5	0.025	0.05	0.10	0.20	0.40
	Low	0.3	0.015	0.03	0.06	0.12	0.24
	Very low	0.1	0.005	0.01	0.02	0.04	0.08
			0.05	**0.1**	**0.2**	**0.4**	**0.8**
			Very low	Low	Medium	High	Very high
			IMPACT				

Key:

High Risk	>0.20:	Review risk in great detail. Amend project strategy to reduce
Medium Risk	0.08 – 0.20:	Develop contingency plans. Monitor risk development
Low risk	<0.08:	Maintain record of risk. Consider contingency measures in outline

to be ranked and compared on an even basis. Where resources are scarce, it also allows attention to be focused on those risks that sit in the high-risk category of the matrix.

(*ii*) *Synthesis of all risks to predict the most likely project outcome.* Risk tends to have a "knock-on" effect, and so it is important to consider not only the effect of individual risks but their cumulative effect as well.

(*iii*) *Investigation of alternative course of actions.* Analysis allows for different scenarios to be developed to provide a high degree of confidence that the appropriate course of action has been chosen. The alternative courses of action may be plotted on the probability/impact matrix to obtain a clearer understanding of their relative merits in relation to the risk profile.

Qualitative Risk Analysis Tools and Techniques. The probability/impact matrix is the easiest technique to use and allows all project team members to participate in the process via a risk workshop. This matrix, shown in Figure 5.5, indicates very quickly those risks that require detailed consideration and those that only need a cursory glance. So, a risk with a high probability of occurring, which does occur, would have a very high impact on the project and would require detailed consideration. This allows the right amount of resource to be used in managing risk.

The probability/impact matrix allows a risk profile for a project to be developed. For instance, if most of the risks are plotted in the top left section of the matrix, it indicates a high-risk project. This matrix is very useful for providing a graphical representation of the risk on a project. The risk management strategy is to manage these risks proactively through the project life cycle. During later risk assessment exercises, the profile should show risks plotted more in the bottom right section of the matrix.

Qualitative risk analysis allows an insight into the risks that a project faces and how those risks can be managed. Many of the tools and techniques allow risks to be scored and

FIGURE 5.5. MATRIX OF PROBABILITY VERSUS IMPACT.

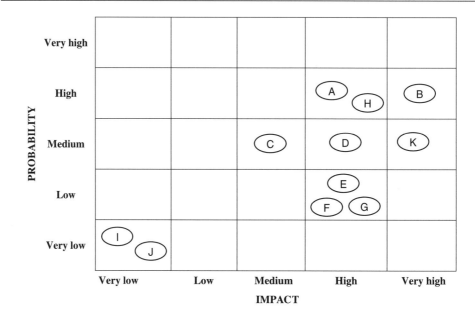

ranked against each other. It should be remembered that the process relies on mainly subjective data. Often much of the data will be a perceptual assessment of a risk provided by project stakeholders.

It can be seen that qualitative risk analysis provides a useful understanding of project risk. The subjective nature of this understanding must not be lost sight of, especially when you are making decisions based solely on qualitative analysis.

A more detailed insight can often be provided by undertaking a quantitative risk analysis.

Quantitative Risk Analysis. Quantitative risk analysis allows project risks to be modeled. Quantitative analysis applies statistical theory to the risk management process. Because of the often complex calculations involved, many of the quantitative analysis techniques are supported by various computer software packages. The UK-based Association for Project Management provides a software directory of available risk management software packages that can be accessed via its Web site.

The use of statistics in quantitative risk analysis tends to enhance the credibility of the output even though there is often inadequacy in the underlying data and assumptions that the model is built upon. For this reason, it is always beneficial to use qualitative techniques to obtain a firm understanding of project risks prior to making the resource intensive investment in the quantitative process. Quantitative techniques should only be used when the case for using them is fully justified.

Decision trees. A *decision tree* is a visual representation of a problem situation and the alternative options that are available for its solution. In Figure 5.6, the project manager must make a decision as to whether the company should stick with an existing contractor who is performing badly or replace the contractor with a new contractor.

The probability of each potential outcome together with its associated cost can be placed on the decision tree. This shows the expected outturn cost of each decision, hence allowing a more informed choice to be made.

Monte Carlo simulation. Monte Carlo simulation generates a number of possible scenarios based on input to the simulation model. Monte Carlo effectively accounts for every possible value that each variable could take and weights each possible scenario by the probability of its occurrence.

The objective of a Monte Carlo simulation is to calculate the combined impact of the model's various uncertainties to determine a probability distribution of the possible model outcomes. The technique involves the random sampling of each probability distribution within the model to produce large numbers of scenarios (often called iterations or trials).

FIGURE 5.6. DECISION TREE.

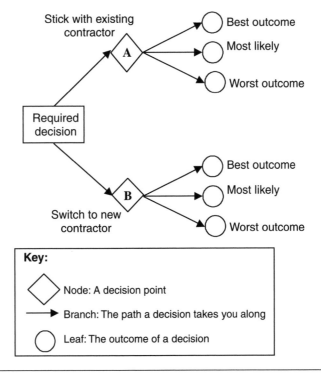

The output from a Monte Carlo simulation is usually displayed as a S curve, as illustrated in Figure 5.7.

Figure 5.7 shows the Monte Carlo output on a cost analysis for a project. This information can be used to set a project budget. The Monte Carlo analysis shows that there is a 50 percent probability the project will cost $229,970. This figure becomes the project budget. A risk allowance can be set at the 60 percent probability level, or $249,180. This would provide the project manager with a risk allowance of $19,210. Of course, the qualitative risk analysis will have identified the risks, so the risk allowance can be spent against known risk events if they occur.

Monte Carlo simulation can also be applied to project schedules. A key output here is the provision of *criticality indexes* for each activity. An activity with a criticality index of 100 percent means that during the simulation, activity was always on the project's critical path. An activity with a criticality index of 75 percent means that activity was on the project's critical path for 75 percent of the simulations. Activities with criticality indexes between 1 and 99 percent are called subcritical. This analysis identifies activities that under certain circumstances may become critical.

Quantitative risk analysis is a complex area that cannot be adequately covered in this introductory chapter. Further guidance is available in Vose (2000), which provides a good overview of quantitative risk analysis and, in particular, its application to project management.

FIGURE 5.7. COST-RISK ANALYSIS S CURVE.

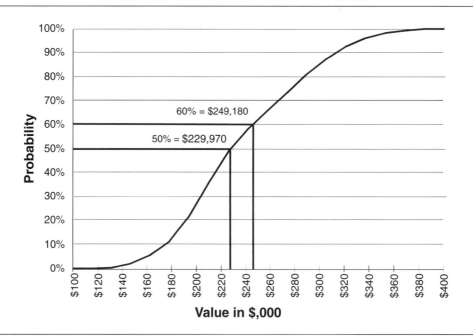

Quantitative analysis provides quite detailed information on the uncertainty that surrounds projects. The modeling process allows the influence of risk to be varied so the impact can be better understood. As with all models, the output is only as good as the input. The byword here is GIGO (garbage in, garbage out).

Risk Response

The analysis stage provides the team with a better understanding of the risks. The next stage is to develop a response to those risks.

Risk Response Methodology. The key concern in this stage of the process is to choose the appropriate course of action. The principle is to choose the right response based on available information. It might be that the response changes with time as more information becomes available. The response will generally fall into one of the following categories:

- *Avoid.* Identifying responses to put in place to sidestep a risk.
- *Transfer.* Transferring a risk from one party to another, for example, from a client to a contractor.
- *Mitigate.* The party who carries a risk should identify responses to lower both the probability of the risk occurring and the impact should it occur.
- *Control.* Responses need to be monitored to ensure they are appropriate in changing circumstances.

The four responses are used in combination with each other, since one will never cover all risk on a project. An important point to consider when developing responses concerns the generation of secondary risk. When a response is proposed, its full implications have to be assessed to ascertain if a secondary risk arises out of implementing the response. If the sum of the secondary risk plus the reduced risk (the original risk with the response in place) is greater than the original risk, an appropriate response has clearly not been identified and an alternative response should be found (see Figure 5.8).

FIGURE 5.8. SECONDARY RISKS.

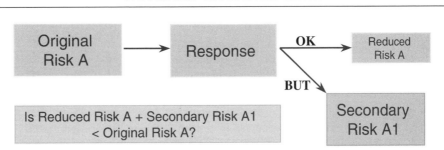

Risk Response Tools and Techniques. The techniques available are primarily integrated with the project management process, for example:

- *Contract acquisition (procurement) plan.* Ensuring that responses to risk are placed into relevant work packages.
- *Contingency management.* Ensuring monies are only released if a predefined event occurs (contingencies are not there to allow for poor budgetary planning).
- *Project controls.* These allow the project to react to changing circumstances, which includes risk.

The tools available are not usually related purely to risk management. For instance, most project managers would recommend to their clients that a contractor has third-party insurance for the project. This is a risk management response designed to reduce clients' exposure to claims for damages if a particular incident were to occur. This is a risk management tool, although most project managers would consider it to be common project procedure. Typical tools available are as follows:

- *Insurances/bonds/warranties.* These are often used to cover the impact of a risk occurring.
- *Contingency plans.* This ensures that should a risk occur, a preprepared plan is put into action;
- *Forms of contracts.* These should be drafted so risks are apportioned as intended by the risk management process;
- *Contingency drawdown models.* Contingency monies should be allocated to specific risks and only released if that risk is within agreed parameters.
- *Special cost allowance.* Not all risk can be foreseen and there is a requirement to allow for unforeseen risks.
- *Training.* This involves ensuring the team understands their role in implementing the risk responses.

All responses to risks that are put in place will need to be managed. This forms part of risk control. Unless responses are effectively managed, the whole risk process fails. Until a response is actioned, the risk may still occur, regardless of how well it has been identified and analyzed.

Risk Control

The whole risk management process relies on there being a control process in place that ensures the risk process is effectively implemented. As part of risk control, a regular review of the risk management process should be undertaken:

- *Risk strategy.* Have any changes to the project been made that would require the risk strategy to be altered?
- *Risk identification.* Have any new risk been identified?

- *Risk assessment.* Assess any new risks that have been identified. Are the assessments for existing risk still valid or should they be revised?
- *Risk response.* Have responses been implemented? Are future responses still valid?

Summary

Risk management should be undertaken as part of a structured, formal process that needs to be aligned to the overall approach to project management. In conclusion, key elements that need to be considered as part of a risk management process are as follows:

- Identify staff and resources assigned to the risk management process.
- Define lines of reporting and responsibility for the risk management process.
- Link the risk management plan to other project tools such as safety, quality and environmental management, and planning and reporting systems.
- Consolidate all risks identified into an appropriate and digestible response strategy in order that cumulative effects can be perceived.
- State risk audit intervals and key milestones.
- Include risk milestones in project plan.
- Identify possible response strategies and programs for each risk category, including contingency plans and how to handle new or unresolved risks.
- Assess cost involved.
- Monitor success of responses strategies, and produce feedback for reporting into future projects.

References and Further Reading

Akintoye, A. 2003. *Public private partnerships: Managing risks and opportunities.* Oxford, UK: Blackwell Science.

British Standards Institute. 2000. *BS 6079—Guide to project management—Part 3 Risk management.* London: British Standards Institute.

Canada Standards Association. 1997. *CAN/CSA-Q850-97 Risk management guidelines for decision makers.* Toronto: Canada Standards Association.

Chapman, C., and S. Ward. 1997. *Project risk management: Processes, techniques, and insights.* Chichester, UK: Wiley.

Chicken, J., and T. Posner. 1998. *The philosophy of risk.* London: Thomas Telford, London.

Godfrey, P. S. 1996. *Control of risk: A guide to the systematic management of risk from construction: SP125.* London: CIRIA.

DoD. 2002. *Risk management guide for DoD acquisition.* 5th ed. Fort Belvoir, VA: Department of Defense, Defense Acquisition University.

ICE. 1998. *RAMP: Risk analysis and management for projects.* London: Institution of Civil Engineers and Institute of Actuaries.

International Organization for Standardization. 2002. *ISO/IEC Guide 73: Risk management—Vocabulary—Guidelines for use in standards.* Geneva: ISO.

Japan Project Management Forum. 2002. *Project management body of knowledge (PMBOK)*. Tokyo: Japan Project Management Forum.

Office of Government Commerce. 2002. *Management of risk: Guidance for practitioners*. London: The Stationery Office.

Project Management Institute. 2000. *Project management body of knowledge (PMBOK)*. Newtown Square, PA: Project Management Institute.

Simon, P., D. Hillson, and K. Newland. 1997. *Project risk analysis and management (PRAM)*. High Wycombe, UK: Association for Project Management.

Smith, N. J. 1998. *Managing risk in construction projects*. Oxford, UK: Blackwell Science.

Standards Australia. 1999. AS/NZS 4360:1999 Risk management. Strathfield, Australia: Standards Association of Australia.

Vose, D. 2000. *Risk analysis: A quantitative guide*. 2nd ed. Chichester, UK: Wiley.

CHAPTER SIX

MAKING RISK MANAGEMENT
MORE EFFECTIVE

Stephen Ward, Chris Chapman

The chapter on risk management by Steve Simister provided an introduction to project risk management and described a generic process for managing risks in projects. This chapter builds on these fundamental ideas to consider how risk management can be made more effective in a given project context. This implies maximizing the benefits obtained from the risk management process for any given level of cost—designing and using processes that are cost-effective.

In designing a risk management application, the project team must address the basic questions associated with the "six Ws" of the risk management process:

- **W**ho wants risk analysis to support risk management, and **w**ho is to undertake it?
- **W**hy is analysis being undertaken?
- **W**hat is the scope of risks to be included in the analysis?
- **W**hichway should the analysis be carried out?
- **W**herewithal—what resources are required?
- **W**hen should analysis be undertaken?

These "six W" questions provide a convenient framework for discussing generic risk management process design principles, although for expository convenience, the *who* question will be considered last. As will become apparent, there are significant interdependencies in the answers to each of these questions, and the order of discussion is somewhat arbitrary when general principles are addressed. For example, several central benefits of risk analysis (*why* issues) depend on how one defines risk, and are not obtainable unless risks are quantified (a *whichway* issue).

115

Why Is Analysis Being Undertaken?

Many people who are not familiar with effective formal risk management assume quite limited roles for risk management, such as simple measurement of risk or identification of things that might go wrong. With this perspective, the benefits of risk management are seen as showing how risky a particular venture is, or protecting project performance from adverse impacts. If this is the limit of the rationale for undertaking risk analysis, then the opportunity to benefit from risk management will be largely wasted. Further, it will prove ineffective in relation to these limited purposes because it will be seen as a counterproductive "add-on" instead of a highly productive "add-in." Effective risk management is not just about measuring or protecting project performance. It is also about assessing and modifying project objectives, base plans, and contingency plans in ways that enhance the prospects for good project performance.

A central reason for employing formal risk management should be to guide and inform the search for favorable alternative courses of action. Central to achieving this is the concept of "risk efficiency," which is concerned with the trade-offs between expected performance and risk that must be made in selecting one course of action or investment strategy over another. Risk efficiency is a core concept in the literature of economics and finance, with a prominence underlined by the award of a Nobel prize for economics to Markowitz (1959) for his portrayal of it in terms of mean-variance portfolio selection models, but the concept is not widely understood by many project managers with engineering backgrounds.

Risk efficiency has to address cost, time, quality, and other measures of performance, but assume, for the moment, that achieved performance can be measured solely in terms of cost outturn, and that achieved success can be measured solely in terms of realized cost relative to some approved cost commitment. In this context risk can be defined in terms of the threat to success posed by a given plan in terms of the size of possible cost overruns and their likelihood. More formally, when assessing a particular project plan in relation to alternatives, we can consider the expected cost of the project as one basic measure of anticipated performance (what should happen on average), and the only other relevant measure of performance as associated cost-risk in terms of downside variability relative to expected cost.

In these terms, some ways of carrying out a project will involve less expected cost and less cost-risk than others—they will be better in both respects, and relatively more efficient. The most efficient plan for any given level of expected cost will involve the minimum feasible level of cost-risk. The most efficient plan for any given level of cost-risk will involve the minimum feasible level of expected cost. Risk efficiency in this sense defines a set of what economists call *Pareto optimal plans*. In choosing between this set of risk-efficient plans, expected cost can only be reduced by adopting a more risky plan, and cost-risk can only be reduced by adopting a plan that increases the expected cost. This concept is most easily pictured using a graph like Figure 6.1.

Consider a set of feasible project plans portrayed in relation to expected cost and cost-risk as indicated in Figure 6.1. The feasible set has an upper and lower bound for both expected cost and cost-risk because there are limits to how good or bad plans can be in both these dimensions.

FIGURE 6.1. RISK-EFFICIENT OPTIONS.

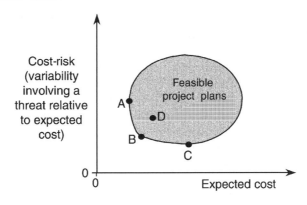

The "risk-efficient boundary" portrayed by the line through points A–B–C defines that set of feasible project plans that provides a minimum level of cost-risk for any given level of expected cost, and the minimum level of expected cost for any given level of cost-risk. Any point off the boundary, like D, represents an inefficient plan, which can be improved upon with respect to both expected cost and cost-risk—moving to B, for example. If a project base plan and associated contingency plans are risk-efficient, any change to these plans that reduces the associated risk will increase the expected cost, and any change that reduces the expected cost will increase the risk.

Diagnosis of potential changes to base or contingency plans to improve risk efficiency is the central purpose of effective project risk management. We can never be sure our plans are risk-efficient. However, we need to search systematically for improvements in risk efficiency, and we need to understand what we are looking for; otherwise, we will never find them. In addition to identifying risk-efficient courses of action, there is a need to consider choices. This involves considering preferred trade-offs between expected cost and level of risk. In relation to Figure 6.1, point A represents the minimum expected cost project plan, with a high level of cost-risk despite its risk efficiency. Choosing A involves not spending money on proactive risk management and a calculated gamble that may not pay off. Point C represents the minimum cost-risk project plan, with a high level of expected cost despite its risk efficiency. Choosing C involves spending money on proactive risk management with the possibility that, in the event of good luck, this expenditure will have been unnecessary.

If an organization can afford to take the risk, A is the preferred solution. Plan A may be the only viable option if an organization's competitors all operate at this minimum expected cost point. Organizations that do not take calculated gambles reduce their average profitability, and this may guarantee going out of business eventually. We have seen programs specifically designed to demonstrate the need for such calculated gambles in organizations that are spending too much on avoiding gambles, the equivalent of persistent overinsurance.

If the risk associated with A is too great, it must be reduced by moving towards C. In general, successive movements will prove less and less cost effective in terms of reducing threat intensity, larger increases in expected cost being required to achieve the same reduction in absolute or relative risk. In practice, an intermediate point like B usually needs to be sought, providing a cost-effective balance between risk and expected cost, the exact point depending upon the organization's ability to take risk.

The scale of the project relative to the organization in question is a key issue in terms of the relative desirability of plans A, B, or C. If the project is one of hundreds, none of which could threaten the organization, plan A may be a sensible choice. If the organization is a one-project organization and failure of the project could lead to failure of the organization, a more prudent stance may be appropriate, adopting a risk-efficient plan closer to C than A. This in turn implies it is very worthwhile defining a level of potential threat below which the organization can ignore cost-risk and above which cost-risk needs to be considered and managed.

In relation to risk-efficient plans and trade-offs between risk and expected performance, risk analysis can help to

- diagnose alternative risk-efficient plans;
- demonstrate the implications of such alternatives; and
- inform choices between alternative risk-efficient plans.

In this way risk management can produce very much more substantial improvements in project performance than a limited focus on merely "keeping things on track." It can also encourage a recognition of the role of considered trade-off decisions involving calculated gambles, so that managers are not blamed for bad luck (and not always held in esteem for experiencing simple good luck).

Taking this perspective, if risk analysis and management is effectively applied to a stream of individual projects over time, a number of important corporate benefits will accrue. Consider, for example, a contracting organization that undertakes risk management prior to and after tendering on individual contracts. For such an organization a number of interrelated benefits can accrue, all driving up profitability, through lower-level benefits like the following:

- *Losing more contracts that ought to be lost.* Better appreciation of uncertainty, which enables more realistic pricing and the avoidance of potential loss making "disaster" contracts where uncertainty is too great.
- *Winning more contracts that ought to be won.* Keener pricing, better design, and stronger risk management abilities providing competitive advantage in terms of improved chances of winning contracts.
- *Lower project costs.* Ability to manage risks to lower project costs with direct profit implications.
- *Lower tendering costs.* Reduce tendering costs through more efficient and effective risk analyses contributing to higher profits directly.

- *More business with higher prices.* Even in the face of substantially lower competitive bids, plausible bids demonstrating effective associated risk management will win when discriminating clients are involved, and these are the most desirable clients.

Risk efficiency, which addresses cost, time, quality, safety, and other measures of performance, may take an aggressive approach to cost-risk, but a conservative approach to safety risk, for example. In any event, the trade-offs between performance criteria and associated uncertainty will need careful management (Klein, 1993). Thoughtful attention to trade-offs and risk efficiency issues can amplify the benefits discussed previously many times over. Conversely, a failure to manage risk efficiency in terms of *all* relevant performance measures can make performance incentives seriously perverse. This is especially important if a contractor is the guardian of issues like quality in the context of incentive contracts (Chapman and Ward, 2002).

In addition to addressing risk efficiency, introducing formal risk analysis and management can lead to valuable cultural changes, treating uncertainty as a source of opportunities rather than as something to be avoided. Formal processes also encourage a proactive approach in which uncertainty is addressed in advance, in a calm and creative way, while there is time to work around problems and exploit opportunities. Reactive crisis management is not eliminated, but it is reduced to a tolerable level. Frustration is reduced, and morale is improved. Such changes can make an organization more exciting to work for and make going to work more enjoyable. This in turn can lead to higher-quality staff wanting to join (and stay with) the organization, with obvious general benefits. All of these benefits should not simply be allowed to happen or not; they should be encouraged by designing them into the risk management process.

It follows that all the preceding benefits are more likely to accrue if organizations develop a corporate capability in (project) risk management, with corporate guidelines and appropriate support, rather than simply encourage the use of risk management in an *ad hoc* way on individual projects (Office of Government Commerce, 2002). It is also the case that organizations with an established infrastructure for supporting risk management will be able to deploy risk management on individual projects much more rapidly and efficiently, and to much greater effect.

What Is the Scope of Risks to Be Included in the Analysis?

A key issue for achieving effective risk management is determining the scope of risks to be included in analysis and subsequent management. Partly this is a question of understanding what is meant by risk in general, and partly it is about delineating the scope of risks that are considered to be project-related. As noted earlier, a process that is limited to a threat perspective on risk management will have limited value because it fails to consider the management of opportunities, in the sense of potential welcome effects on project performance.

In any given decision situation, both threats and opportunities are usually involved, and both should be managed. A focus on one should never be allowed to eliminate concern for the other. While opportunities and threats can sometimes be treated separately, they are seldom independent—just as two sides of the same coin can be examined one at a time, but they are not independent when it comes to tossing the coin. Just as it is inadvisable to pursue opportunities without regard for the associated threats, it is rarely advisable to concentrate on reducing threats without considering associated opportunities. Courses of action are often available that reduce or neutralize potential threats and simultaneously offer opportunities for positive improvements in performance.

To emphasize the desirability of a balanced approach to opportunity and threat management, the term *uncertainty management* is increasingly used in preference to *risk management* and *opportunity management*. However, uncertainty management is not just about managing perceived threats, opportunities, and their implications. It is also about identifying and managing all the sources of uncertainty that give rise to and shape our perceptions of threats and opportunities. It involves exploring and understanding the origins of project uncertainty before seeking to manage it, with no preconceptions about what is desirable or undesirable. Key concerns are understanding where and why uncertainty is important in a given project context, and where it is not. This is a significant change in emphasis compared with most project risk management processes.

Uncertainty in the plain-English sense of lack of certainty is in part about *variability* in relation to performance measures like cost, duration, or quality. It is also about *ambiguity* associated with lack of clarity because of the behavior of relevant project players, lack of data, lack of detail, lack of structure to consider issues, restrictive working and framing assumptions used to consider the issues, known and unknown sources of bias, and ignorance about how much effort it is worth expending to clarify the situation. These aspects of uncertainty can be present throughout the project life cycle, but they are particularly evident during conception, design, and planning, and they contribute to uncertainty in five areas:

- The variability associated with estimates of project parameters
- The basis of estimates of project parameters
- Design and logistics
- Objectives and priorities
- Relationships between project parties

All these areas of uncertainty are important, but generally areas are more fundamental to project performance as we go down the list. Potential for variability is the obvious focus of the first area, but ambiguity rather than variability becomes the more dominant underlying issue in latter areas. Uncertainty about variability associated with estimates is often driven by the other four areas, each of them in turn involving dependencies on later areas in this list.

Variability Associated with Estimates

An obvious area of uncertainty is the size of project parameters such as time, cost, and quality related to particular activities. For example, we may not know how much time and effort will be required to complete a particular activity. The causes of this uncertainty might include one or more of the following:

- Lack of a clear specification of what is required
- Novelty in terms of a lack of experience of this particular activity
- Complexity in terms of the number of influencing factors and interdependencies between these factors
- Limited analysis of the processes involved in the activity
- Possible occurrence of particular events or conditions that could have some (uncertain) effect on the activity

Only the last of these items really relates to specific events or conditions that might be thought of as threats or opportunities. The other sources of uncertainty arise from a lack of understanding of what is involved. Because they are less obviously described as threats or opportunities, they may be missed unless a broad view of uncertainty management is adopted.

Uncertainty about Assumptions Underlying Estimates

A particularly important source of uncertainty is the nature of assumptions underpinning estimates. The need to note assumptions about resources, choices, and methods of working is well understood if not always fully operationalized. However, a large proportion of those using probabilistic project risk management processes often fail to address the conditional nature of probabilities and associated measures used for decision making and control. Key outputs of estimation and evaluation phases of the risk management process (see the chapter by Simister) are estimates of expected values for project parameters and measures of plausible variations on the high and low side. Interpretation of expected values, or plausible extremes like a 95 percent confidence value, have to be conditional on the assumptions made to estimate these values. For example, a sales estimate may be conditional on a whole set of assumed trading conditions, such as a particular promotion campaign and no new competitors. Invariably, estimates ignore, or assume away, the existence of uncertainty, which relates to three basic sources: known unknowns, unknown unknowns, and bias:

- *Known unknowns* are of two types: explicit, extreme events (triple Es), and scope adjustment provisions (SAPs). Triple Es are *force majeure* events, like a change in legislation that would influence an oil company's pipeline design criteria in a fundamental way. SAPs are conditions or assumptions that may not hold and that are explicit, like the assumed operating pressure and flow value for an oil pipeline, given the assumed oil recovery rate.
- Unknown unknowns are the unidentified triple Es or SAPs that should be factored in to the risk management process. We know that the realization of some unknown unknowns is usually inevitable. They do not include issues like "the world may end tomorrow," because it is sensible for most practical decision making to assume we will still be here tomorrow, but the boundary between this extreme and what should be included is usually ill-defined.

- Bias may be conscious or unconscious, pessimistic or optimistic (McCray, Purvis and McCray, 2002). It is usually difficult to identify, and clues or data may be available or not.

All three of these sources of uncertainty can have a very substantial impact on estimates, and this needs to be recognized and managed.

Uncertainty about Design and Logistics

In the conception stage of the project life cycle the nature of the project deliverable and the process for producing it are fundamental uncertainties. In principle, most of this uncertainty should be removed in pre-execution stages of the life cycle by specifying what is to be done; how, when, and by whom; and at what cost. In practice, a significant amount of this uncertainty may remain unresolved through much of the project life cycle (see, for example, Drummond, 1999). The nature of design and logistics assumptions and associated uncertainty may drive some of the uncertainty about the basis of estimates.

Uncertainty about Objectives and Priorities

Major difficulties arise in projects if there is uncertainty about project objectives, the relative priorities between objectives, and acceptable trade-offs. These difficulties are compounded if this uncertainty extends to the objectives and motives of the different project parties, and the trade-offs parties are prepared to make between their objectives. A key question is "Do all parties understand their responsibilities and the expectations of other parties in clearly defined terms that link objectives to planned activities?" Value management has been introduced to encompass this concern (Kelly and Male, 1993). The need to do so is perhaps indicative of a perceived failure of risk management practices. However approached, *risk management and value management need joint integration into project management.*

Uncertainty about Fundamental Relationships between Project Parties

Uncertainty about objectives and priorities is compounded by any uncertainty about the identity of project parties, their respective roles, and their relationships with one another. The relationships between the various parties may be complex. They may, or may not, involve formal contracts. The involvement of multiple parties in a project introduces uncertainty arising from ambiguity with respect to the following:

- Specification of responsibilities
- Perceptions of roles and responsibilities
- Communication across interfaces
- The capability of parties
- Contractual conditions and their effects
- Mechanisms for coordination and control

Ambiguity about roles and responsibilities for bearing and managing project-related uncertainty may be involved here. For example, interpretations of risk apportionment implied by standard contract clauses may differ between contracting parties (Hartman and Snelgrove, 1996; Hartman, Snelgrove, and Ashfrati, 1997). This ambiguity ought to be addressed systematically in any project, not just in those involving formal contracts between different organizations. Contractor organizations are often more aware of this source of ambiguity than their clients, although the full scope of the risks and opportunities that this ambiguity generates for each party in any contract (via claims, for example) may not always be fully appreciated until rather late in the day (Ackermann et al., 1997; Cooper, 1980; Williams et al., 1995).

Whichway Should Analysis Be Carried Out

An important aspect of developing cost-effective approaches to risk management involves consideration of the appropriate structure and level of detail in the analysis to be undertaken, and the choice of models and techniques. Choices here will be strongly influenced by the who?, why?, and what? questions discussed previously.

To address uncertainty in both variability and ambiguity terms, we need to modify and augment existing risk management processes and adopt a more explicit focus on uncertainty management. An obvious first step is to replace terminology involving the word *risk* with the word *uncertainty* and avoid using purely threat-orientated descriptors in identification exercises. Other steps involve modifications to risk management processes to address each of the sources of uncertainty outlined in the previous section. What these modifications involve is outlined in the following text. More detailed explanations and commentary can be found in Chapman and Ward (2002).

Expose and Investigate Variability

Difficulty in estimating time or effort required to complete a particular activity may arise from a lack of knowledge of what is involved rather than from the uncertain consequences of potential threats or opportunities. Attempting to address this difficulty in conventional risk management terms is not appropriate. What is needed is action to improve knowledge of organizational capabilities and reduce variability in the performance of particular project-related tasks. For example, uncertainty about the time and cost needed to complete design or fabrication in a project may not be readily attributable to particular sources of risk, but to variability in efficiency and effectiveness of working practices. An uncertainty management perspective would seek an understanding of why this variability arises, with a view to managing it. This may require going beyond addressing uncertainty associated with a specific project, to trigger studies of operations that provide an input into a range of projects, as illustrated by this example.

Further investigation of variability, and consideration of risk-efficient alternative courses of action in base plans or contingency plans, requires quantification of the perceived variability. Single-point estimates of a particular parameter are of limited value for uncertainty management purposes without some indication of the potential variability in the size of the parameter. For example, a best estimate of the cost of a particular activity is of limited value without some indication of the range or probability distribution of possible costs. Quantifying variability (and uncertainty about this variability) forces management to articulate beliefs about uncertainty and related assumptions. In addition, quantification obliges managers to clarify the significance of differences between *targets, expected values* and *commitments*, with respect to costs, durations, and other performance measures. This in turn highlights the need to clarify the distinction between *provisions* and *contingency allowances*.

In cost terms, expected values are our best estimate of what costs should be realized on average. Setting aside a contingency fund to meet costs that may arise in excess of the expected cost defines a probability of being able to meet the commitment. The contingency allowance provides an uplift from the expected value, which is not required on average if it is properly determined. Determining this probability of being able to meet a commitment ought to involve an assessment of all related downside variations and the extent to which these may be covered by a contingency fund, together with an assessment of the implications of both over- and underachievement in relation to the commitment.

Targets, set at a level below expected cost, with provisions accounting for the difference, need to reflect the opportunity aspect of risk. Targets need to be realistic to be credible, but they also need to be lean, to stretch people. If optimistic targets are not aimed for, expected costs will not be achieved on average, and contingency funds will be used more often than anticipated.

Organizations that do not quantify risks have no real basis for distinguishing these three very different kinds of estimates. As a consequence, single values attempt to serve all three purposes, usually with obviously disastrous results, not to mention costly and unnecessary dysfunctional organizational behavior. The *cost estimate*, the *completion date*, or the *promised performance* become less and less plausible, there is a crisis of confidence when they are moved, and then the process starts all over again. Sometimes differences between targets, expectations and commitments are kept confidential, or left implicit. Effective risk management requires these differences to be explicit, and a clear rationale for the difference needs to be understood by all, leading to an effective process of managing the evolution from targets to realized values. The ability to manage the gaps between targets, expected values, and contingency levels, and setting those values appropriately in the first place, is a central concern of risk management.

Quantifying uncertainty about levels of performance in terms of targets, expectations, and commitments is useful if concern is with aggregate performance on a single criterion such as cost or time. However, this approach is less helpful if applied to quantifying uncertainty about each activity individually in a chain or collection of activities. For example, setting commitment levels for the duration of each task in a chain of tasks may be counterproductive. In *Critical Chain* Goldratt (1997) describes the problem in the following terms: "we are accustomed to believing that the only way to protect the whole is through protecting the completion date of each step"; as a result, "we pad each step with a lot of safety time." Goldratt argues that this threat protection perspective induces three behaviors that, when combined, waste most of the safety time:

- The student syndrome (leaving things to the last minute)
- Multitasking (chopping and changing between different jobs)
- Delays accumulate, advances do not (good luck is not passed on)

The challenge, and opportunity, is to manage the *uncertainty* about performance for each link in the chain in a way that avoids these effects and ensures that good luck is not only captured but shared for the benefit of the whole project. Joint management of the good luck, efficiency, and effectiveness of each project-related activity is needed. This implies some form of incentive agreement between the project manager and those carrying out project activities that encourages the generation and delivery of good luck, efficiency, and effectiveness, with the minimum of uncertainty. This agreement needs to recognize interdependency between performance measures. Duration, cost, quality, and time are not independent, and all four are functions of the motivation and priorities of those involved. For example, uncertainty about the duration of an activity is usually driven by ambiguity about quality, cost, and inefficient working practices. This needs to be managed, to reduce uncertainty and to capture the benefits of managing good luck, in the sense of the scope for low duration, high quality, and low cost.

To illustrate these ideas briefly, consider a procurement project involving significant design work before the next stage of the project can proceed. If an internal design department is involved, do we need to set a commitment date for design completion? The simple answer is we do not. Rather, we need a target duration plus an expected duration that becomes firm as early as possible, in order to manage the good luck as well as the bad luck associated with variations in the duration of the design activity. We also need an agreement with the design department that recognizes that the design department can make and share their luck to a significant extent if they are motivated to do so. As the design department is part of the project-owning organization, a legal contract is not appropriate. However, a contract is still needed in the form of a memorandum of understanding, to formalize the agreement. Failure to formalize an internal contract in a context like this implies psychological contracts between project parties that are unlikely to be effective. The ambiguity inherent in such contracts can only generate uncertainty, which is highly divisive and quite unnecessary.

Most internal design departments have a *cost per design hour* rate based on an historic accounting cost. A *design hours* estimate times this rate yields a design cost estimate. Design actual cost is based on realized design hours and the internal contract is *cost plus*. The duration agreed to by the design department is a "commitment" date with a low chance of being exceeded, as noted earlier. To address the problems that this arrangement induces, some form of incentive agreement is required that recognizes the potential for trade-offs between different measures of performance. What is needed is a *fixed nominal cost* based on the appropriate expected number of design hours, with a premium payment scale for completion earlier than an appropriate *trigger duration*, and a penalty deduction scale for later completion. Additionally, a premium could be introduced for the correct prediction of the design completion date to facilitate the more efficient preparation of following activities. The trigger duration might be something like an 80-percentile value, comparable to a commitment duration. The target should be very ambitious, reflecting a plausible date if all goes as well as possible. Other performance objectives can be treated in the same way, with premium and penalty payments relative to cost and quality level triggers as possible options. However, premiums and penalties need to be designed to ensure that appropriate trade-offs are encouraged.

Such *trigger contracts* can have beneficial effects on performance that go beyond the immediate project. For example, carrying on with the preceding example, if a trigger-based incentive contract with an internal design department is used, fewer hours may be required because the incentive structure will reduce multitasking. If multitasking is reduced, the efficiency of the design department might improve enough to eliminate rumors of selling off the design function (outsourcing all design), which might be underpinning risks related to loss of staff and low morale. If these downside risks are eliminated, improved morale, low turnover of good staff, and easier hiring may follow leading to further gains in efficiency and effectiveness in a virtuous circle.

Clarify Assumptions Underlying Estimates

The previous section noted that estimates of project parameters invariably ignore, or assume away, the existence of uncertainty that relates to three basic sources: known unknowns, unknown unknowns, and bias. A full description of how best to address these sources of uncertainty is complex. An outline is provided elsewhere (Chapman and Ward, 2002), a summary here.

The starting point is recognizing that all estimates are *conditional*, in the sense that assumptions they depend upon have been made. The second step is understanding that all estimates are *subjective*, in the sense that truly objective estimates that are fit for purpose do not exist. The third step is recognizing that judging the quality of someone else's estimates necessarily involves judging the quality of the process they used to arrive at that estimate, as well as the nature of their explicit and implicit conditions.

The impact of these three basic sources of uncertainty can be considered via the use of three scaling factors: F_k known unknowns, F_u unknown unknowns, and F_b bias. For example, an F_k scaling factor might be defined by the user of an estimate in relation to an estimated expected cost provided by a provider of the estimate in the form:

$$F_k = 1.0 \qquad \text{Probability of } F_k = 0.1$$

$$1.1 \qquad\qquad\qquad 0.7$$

$$1.2 \qquad\qquad\qquad 0.1$$

$$1.3 \qquad\qquad\qquad 0.1$$

This example involves a mean $F_k = 1.0 \times 0.1 + 1.1 \times 0.7 + 1.2 \times 0.1 + 1.3 \times 0.1 = 1.12$. The subjective probabilities might be based on the user's view of the impact of conditions noted by the provider, of the type "normal market conditions will be involved."

In simple, crude terms, the example numbers would imply that an uplift in the estimated cost of the order of 30 percent is plausible for a pessimistic scenario value, like a 95 percentile, and an uplift of 12 percent is an appropriate expectation, because of the potential impact of an abnormal market or the violation of other conditions noted. In practice, F_k values could be much higher than in this example. A very careful risk-management-driven estimation process resulting in a $F_k = 1$ might suggest an $F_u = 1$ and an $F_b = 1$, but much higher values may be appropriate.

Combining these three scaling factors provides a single *cube factor* (short for kuuub from **k**nown **u**nknowns, **u**nknown **u**knowns, and **b**ias), designated F^3, and defined by $F^3 = F_k \times F_u \times F_b$, which is then applied as scaling factor to conditional estimates. This cube factor, F^3, can be estimated in probability terms directly or via these three components to clarify the conditional nature of the output of any quantitative risk analysis. This avoids the very difficult mental gymnastics associated with trying to interpret a quantitative risk analysis result that is conditional on exclusions and scope assumptions (which may be explicit or implicit), and no bias, without underestimating the importance of the conditions.

The key value of explicit quantification of F^3 is forcing those involved to think about the implications of the factors that drive the expected size and variability of F^3. Such factors may be far more important than the factors captured in the prior conventional quantitative risk analysis. There is a natural tendency to forget about conditions and assumptions and focus on the numbers. Attempting to explicitly size F^3 makes it possible to avoid this. Even if different parties emerge with different views of an appropriate F^3, the process of discussion is beneficial. If an organization refuses to estimate F^3 explicitly, the issues involved do not go away; they simply become unmanaged risks. Many of them will be betting certainties. Variability, that needs to be managed in a risk management context, must embrace cube factors explicitly, and "variability" is defined here in this sense.

An important source of ambiguity concerns the extent to which different project parties need to be concerned about particular cube factors. For example, suppose a project manager decides to contract out design work. In estimating design costs, the design contractor will not scale its estimates to allow for known unknowns if the contractor can negotiate a contract to avoid bearing any risk associated with known unknowns. Similarly, it will not be appropriate for the project manager to scale the project design budget to incorporate an allowance for these known unknowns, unless they are wholly under the control of the project manager. For example, certain scope adjustments may come in this category. The potential impact of other unknown unknowns needs to be recognized at some organizational level above the project manager, where there is an ability to bear the consequences of any unknown unknown occurring, and an appropriate cube factor estimated. A similar cube factor would need estimation if the project company's own design department undertook the work, but the risk allocation issues would be more complicated. Much post project litigation arises because of a failure to appreciate or acknowledge exposure to cube factors, and a failure to resolve ambiguity about responsibility for cube factors earlier in the project.

Address Uncertainty about Fundamental Relationships, as well as Design and Logistics

Careful attention to formal risk management is usually motivated by the large-scale use of new and untried technology while executing major projects, where there are likely to be significant threats to achieving objectives. A threat perspective encourages a focus on these initial motivating risks. However, key issues are often unrelated to the motivating risks and are usually related to sources of ambiguity introduced by the existence of multiple parties and the project management infrastructure. Such issues need to be addressed very early in the project and throughout the project life cycle, and should be informed by a broad appreciation of the underlying roots of uncertainty. A decade ago most project management professionals would have seen this in terms of a suitable high-level activity structure summary and a related cost item structure. It is now clear that an activity structure is only one of six aspects of a project that need consideration (Chapman and Ward, 2003). To review, these are as follows:

- Who (parties or players involved)
- Why (motives, aims, and objectives of the parties)
- What (design of the deliverable)
- Whichway (activities to achieve the deliverable)
- Wherewithal (resources required)
- When (time frame involved)

Understanding the sources of uncertainty associated with each of these aspects is fundamental to effective identification and management of both threats and opportunities. Use of this "six Ws" framework for the project from the earliest stages of the PLC could usefully inform development of project design and logistics by clarifying key sources of uncertainty. However, it is important not to treat all these sources of uncertainty as independent—an assumption that is implicitly encouraged by summary risk registers (Williams, 2000). In practice it is better to assume dependency and seek to understand its nature, recognizing that dependencies can be complex, involving chains of knock-on effects and undesirable self-reinforcing feedback loops (see, for example, Drummond, 1999; Eden et al., 2000). Such dependencies can be effectively represented using influence diagrams, causal maps, and systems dynamics models (Ashley and Avots, 1984; Eden et al., 2000; Howick and Eden, 2001).

Address Uncertainty about Objectives and Priorities

As part of the process of understanding the relationships between the six Ws of the project, project owners need to

1. identify pertinent performance criteria;
2. develop a measure of the level of performance for each criterion;
3. identify the most preferred (optimum) feasible combination of performance levels on each criterion;
4. identify alternative combinations of performance levels on each criterion that would be acceptable instead of the "optimum"; and
5. identify the trade-offs between performance criteria implied by these preferences.

These steps should be undertaken by any project owner, particularly in the early conception and design stages of a project. Adopting an iterative process may be the most effective way to complete these steps. The process of identifying and considering possible trade-offs between performance criteria is an opportunity to improve performance, as noted earlier. It should enable a degree of optimization with respect to each performance measure, and it is an opportunity that needs to be seized. In particular, the information gathered from these steps can be used to formulate appropriate incentive contracts by selecting some or all of the performance criteria for inclusion in the contract, developing payment scales that reflect the acceptable trade-offs, and with the supplier, negotiating acceptable risk-sharing ratios for each contract performance criterion. A detailed discussion on the formulation of such contracts is given in Chapman and Ward (2002, Chapter 5).

Constructively Simple Estimating

To facilitate insight and learning in a cost-effective process, uncertainty in all the various forms discussed previously needs to be explored using an iterative process, with process objectives that change on successive passes. An iterative approach is essential to optimize the use of time and other resources during the uncertainty management process, because initially we do not know where uncertainty lies, whether or not it matters, or how best to respond to it. To begin with, a first pass is usually about sizing variability, to see if it might matter, and to reflect potential variability in an unbiased estimate of the expected outcome. If a very simple first-pass conservative estimate suggests expected values and variability do not matter, we should be able to ignore variability without further concern or further effort. If the first pass raises concerns, further passes are necessary in order to effectively manage what matters. Final passes may be concerned with convincing others that what matters is being properly managed. The way successive iterations are used needs to be addressed in a systematic manner. A simple one-shot, linear approach is hopelessly inefficient.

A common first-pass approach to estimation and evaluation employs a probability-impact matrix (PIM). The PIM approach typically defines low, medium, and high bands for possible probabilities and impacts associated with identified sources of uncertainty (usually risks involving adverse impacts). These bands may be defined as quantified ranges or left wholly subjective. *The PIM approach offers a rapid first-pass assessment of the relative importance of identified sources of uncertainty, but otherwise delivers very little useful information* (Ward, 1999a).

Even with the availability of proprietary software products such as Risk for quantifying, displaying, and combining uncertain parameters, use of PIM has persisted (further encouraged by PIM software). This is surprising, but it suggests a gap between simple direct prioritization of sources of risk and quantification requiring the use of specialist software. To address this gap, Chapman and Ward (2000) describe a 'minimalist' first-pass approach to estimation and evaluation of uncertainty. This approach defines uncertainty ranges for probability and impact associated with each source of uncertainty. Subsequent calculations preserve expected value and measures of variability, while explicitly managing associated optimistic bias.

The minimalist approach involves the following steps in a first-pass attempt to estimate and evaluate uncertainty:

1. Identify the parameters to be quantified.
2. Estimate crude but credible ranges for probability of occurrence and impact.
3. Calculate expected values and ranges for composite parameters.
4. Present results graphically (optional).
5. Summarize results.

In step 1 a clear distinction is made between sources of uncertainty that are useful to quantify and sources that are best treated as possible scenarios. For example, suppose an oil company project team wants to estimate the duration and the cost of the design of an offshore pipeline using the organization's own design department. Current common best practice would require a list of sources of uncertainty (a risk list or risk log), which might include entries like "change of route," "demand for design effort from other projects," "loss of staff," and "morale problems". "Changes of route" would probably be regarded as a source best treated as a condition by the project manager and by the head of the design department. Subsequent steps apply only to those sources of uncertainty that are usefully quantified.

In step 2 the probability of a threat occurring is associated with an approximate order of magnitude minimum and maximum plausible probability, assuming a uniform distribution (and a midpoint expected value). This captures the user's feel for a low, medium, or high probability class in a flexible manner, captures information about uncertainty associated with the probability, and yields a conservative (pessimistic) expected value. For trained users it should be easier than designing appropriate standard classes for all risks and putting a tick in an appropriate box.

Similarly, in step 3 the impact of a threat that occurs is associated with an approximate order of magnitude minimum and maximum plausible value (duration and cost), also assuming a uniform duration. This captures the user's feel for a low, medium, or high impact class in a flexible manner, captures information about the uncertainty associated with the impact, and yields a conservative (pessimistic) expected value. For trained users it should be easier than designing appropriate standard classes for all risks and putting a tick in an appropriate box.

In step 4 the expected values and associated uncertainties for all quantified sources of uncertainty are shown graphically in a way that displays the contribution of each to the total, in expected value and range terms, clearly indicating what matters and what does not, as a basis for managing subsequent passes of the risk management process in terms of data acquisition to confirm important probability and impact assessment, refinement of response strategies, and key decision choices.

Although simple, the minimalist approach is sophisticated in the sense that it builds in pessimistic bias to minimize the risk of dismissing as unimportant risks that more information might reveal as important. Also, it is set in the context of an iterative approach that leads to more refined estimates wherever potentially important risks are revealed. Sophistication does not require complexity. It requires *constructive simplicity*, increasing complexity only when it is useful to do so.

The concern of the minimalist first-pass approach is not a defensible quantitative assessment. The concern is to develop a clear understanding of what seems to matter based on the views of those able to throw some light on the issues. This is an attempt to resolve the ambiguity associated with the size of uncertainty about the impact of risk events and the size of uncertainty about the probability of risk events occurring, the latter often dominating the former. A first pass may lead to the conclusion there is no significant uncertainty, and no need for further effort. This is one of the reasons why the approach must have a conservative bias. Another reason is the need to manage expectations, with subsequent refinements of estimates indicating less uncertainty/more uncertainty providing an explicit indication that the earlier process failed. An estimator should be confident that more work on refining the analysis is at least as likely to decrease the expected value estimate as to increase it. A tendency for cost estimates to drift upward as more analysis is undertaken indicates a failure of earlier analysis. The minimalist approach is designed to help manage the expectations of those the estimator reports to in terms of expected values. Preserving credibility should be an important concern.

Readers used to single-pass approaches that attempt considerable precision may feel uncomfortable with the deliberate lack of precision incorporated in the minimalist approach. However, more precise modeling is frequently accompanied by questionable underlying assumptions such as independence between parameters and lack of attention to uncertainty in original estimates. The minimalist approach forces explicit consideration of these issues.

What Resources (Wherewithal) Are Required?

Just as resources for the project require explicit consideration, so too do resources for effective risk analysis and management, the process *wherewithal* question. In a given project context, there may be specific constraints on cost and time. Resource questions are likely to revolve around the availability and quality of human resources, including the availability of key project personnel, and the availability of information processing facilities.

If one of our clients asks "How long will it take to assess my project's risk?," the quite truthful response "How long is a piece of string?" will not do. A more useful response is "How long have we got?" (the process *when*), in conjunction with "How much effort can be made available?" (the process *wherewithal*), "Who wants it?", (the process *who*), and "What do you want it for?" (the process *why*). The answer to this latter question often drives the process *what* and the *whichway*. It is important to understand the interdependence of these considerations. Six months or more may be an appropriate duration for the initial, detailed project risk analyses of a major project, but six hours can be put to very effective use if the question of the time available is addressed effectively in relation to the other process *W*s. Even a few minutes may prove useful for small projects. In these circumstances, it is essential to adopt an iterative approach with a minimalist first pass as outlined in the previous section.

Computing power is no longer a significant constraint for most project planning, with or without consideration of uncertainty. Even very small projects can afford access to powerful personal computers. However, software can be a significant constraint, even for very large projects. It is important to select software that is efficient and effective for an appropriate model and method. It is also important to prevent preselected software from unduly shaping the form of the analysis.

In the early stages of the risk management process, the risk analysis team may be seen as the project planning player doing most of the risk management running. However, it is vital that all the other players see themselves as part of the team and push the development of the risk management process as a vehicle serving their needs. This implies commitment and a willingness to spend time providing input to the risk analysis and exploring the implications of its output.

When Should Analysis Be Undertaken?

The nature of projects undertaken is likely to be a primary influence on the scope, level of detail, perspective, and extent of quantification that is appropriate. For example, a combination of substantial investment with high levels of uncertainty warrants serious management attention, greatly facilitated by a comprehensive, systematic risk management process. However, if projects are of a routine, low-risk nature, project managers hardly need sophisticated risk management systems.

In general, risk management could be usefully applied on separate and different bases in each stage of the project life cycle without the necessity for risk management in any previous or subsequent stages. For example, risk analysis could form part of an "evaluation" step in any stage of the life cycle. Alternatively, risk analysis might be used to guide initial progress in each life cycle stage. In these circumstances, the focus of risk analysis is likely to reflect immediate project management concerns in the associated project stage. For example, risk analysis might be undertaken as part of the Plan stage primarily to consider the feasibility and development of the work schedule for project execution. There might be no expectation that such risk analysis would or should influence the design, although it might be perceived as a potential influence on the subsequent work allocation decisions. In practice, many risk analyses are intentionally limited in scope, as in individual studies to determine the reliability of available equipment, to determine the likely outcome of a particular course of action, or to evaluate alternative decision options within a particular project life cycle stage. This can be unfortunate if it implies a limited, *ad hoc* bolted-on or optional extra approach to risk management, rather than undertaking risk management as an integral built-in part of project management.

As a general rule, the earlier risk management can be carried out in the project life cycle, the better. Implementing risk management earlier than the planning stage can be difficult because the project is more fluid and less well defined, with more degrees of freedom and more alternatives to consider. A less well defined project also means appropriate documentation is harder to come by and alternative interpretations of what is involved may not be resolvable (Uher and Toakley, 1999). That said, implementing risk management earlier in the life cycle is usually much more useful than if it is first attempted later on. Risk management earlier in the life cycle is usually less quantitative, less formal, less tactical, more strategic, more creative, and more concerned with the identification and capture of opportunities. There is scope for much more fundamental improvements in project plans, perhaps including initial design or redesign of the product of the project. Also, it can be particularly useful to be very clear about project objectives as early as possible, in the limit-decomposing-projects objectives and formally mapping their relationships with project activities, because preemptive responses to threats need to facilitate lateral thinking that addresses entirely new ways of achieving objectives.

Implementing risk management later in the project life cycle gives rise to somewhat different difficulties, without any compensating benefits. Contracts are in place, equipment has been purchased, commitments are in place, reputations are on the line, and managing change is comparatively difficult and unrewarding. Risk management can and should encompass routine reappraisal of a project's viability. In this context, early warnings are preferable to late recognition that targets are incompatible or unachievable, but better late than never.

Who Wants the Risk Analysis, and Who Is to Undertake It?

In any given project context, an important initial step in scoping the project risk management process is to clarify *who* is undertaking risk analysis for whom, and how the reporting process will be managed. The key players should be as follows:

- Senior managers, to empower the process, to ensure the risk analysis effort reflects the needs and concerns of senior managers, and to ensure it contains the relevant judgments and expertise of senior managers
- All other relevant managers, to make it part of the total management process and ensure that it services the whole project management process
- All relevant technical experts, to ensure it captures all relevant expertise for communication to all relevant users of that expertise in an appropriate manner
- A risk analyst or risk analysis team, to provide facilitation/elicitation skills, modeling and method design skills, computation skills, and teaching skills that get the relevant messages to all other members of the organization and allow the risk analysis function to develop and evolve in a way that suits the organization.

Some organizations refer to risk analysts as risk *managers*. This usually implies a confusion of roles. Risk is a pervasive aspect of a project that can be delegated in terms of analysis, but not in terms of management. Proper integration of project risk management and project management requires that the project manager takes personal responsibility for all risk not explicitly delegated to managers of components of the project. Further, ownership of the risk management process should not be portioned off. It needs to be embedded in the thought processes and actions of the team as a whole.

Drivers of Participant Performance

As with any task, the effectiveness of a project participant in undertaking risk management is driven by both the working environment and characteristics of the participant (Ward, 1999b). From a risk management perspective, the working environment could be characterized by factors such as location of the risk management effort in relation to the project organization; the nature of the organizational structure; the quality of supporting information systems; the availability of information; and the existence of an organization culture with respect to attitude to risk and risk management, resources available, and time available to undertake risk management. These factors either have to be managed to facilitate risk management or else taken into account in determining the form of risk management process that is feasible.

In terms of effectively undertaking risk management, relevant characteristics of a project participant comprise the participant's perception of their responsibilities for undertaking risk management in the given project, the participant's capability and experience in risk management, and his or her motivation to undertake risk management.

Perceived Responsibilities

The need for clearly specified responsibilities has long been recognized as a central requirement for effective performance in organizations, reflected in the widespread use of "management by objectives" and related techniques. In a project context this translates into a need for responsibility for project tasks to be clearly allocated to one or more parties, and for these parties to have a clear idea of what is expected of them. In terms of risk management, this translates into a need for sources of uncertainty to be identified and clearly allocated to appropriate parties. For example, a client organization may seek to manage risks by implicitly transferring them to a contractor via a firm fixed-price contract, but this is no guarantee that the contractor will identify these risks and manage them in the client's interest.

Capability and Experience

In any project, a participant's perception of his or her responsibilities will depend on the nature of the project being undertaken and the extent of their capability and experience. This applies to project tasks in general and risk management in particular. For example, contractual requirements that a contractor shall be responsible for certain project risks, or undertake risk management, are no guarantee that effective risk management will actually occur. Selecting contractors with appropriate capability and experience in risk management will go some way to ensuring that such contractual obligations are met to the satisfaction of the client.

Similar considerations apply to project parties within the project owner's own organization. In any given project, the risk analysis team must be seen by the project team as an effective support function. If this is not the case, the cooperation necessary to do the job will not be forthcoming and the risk management process will flounder. However, it is equally important that the risk analysis team be seen by the project owning organization as unbiased, with a demonstrable record for telling it as it is, as providers of an honest broker external review as part of the process. If this is not the case, the risk management process will sink without trace.

Undertaking risk management is a high-risk project in itself, especially if embedding effective risk management in the organization as well as in the project in question is the objective. Often the project planning team provide a high-risk environment for risk analysis because, for example, project management is ineffective, or project team members

- are not familiar with effective project risk management processes;
- are familiar with inappropriate risk management processes;
- come from very difficult cultures;
- come from competing organizations or departments.

If the quality of the project management process or staff is a serious issue, this can be the biggest threat to the risk management process, as well as to the project itself. If this is the case, it deserves careful management, for obvious reasons.

Recognition of these issues by top management usually results in initiatives to enhance *corporate* capability in project risk management. Hillson (1997) has characterized corporate capability for risk management in terms of four levels of risk maturity:

1. The level of commitment to risk management
2. The degree of formality in risk management processes
3. The level of in-house expertise and training in risk management skills
4. The extent to which risk management tools and methods are applied to the organization's activities

Once an organization has assessed its level of risk maturity in terms of these four attributes, Hillson identifies a number of actions that the organization then needs to undertake in order to raise its level of risk maturity.

Motivation

With respect to risk management, project participants need to be convinced that risk management activity will help them meet their own objectives. For powerful customers, a requirement that all tenders include a risk management plan may be sufficient inducement for offering contractors to comply at some level. However, if the customer ignores the needs of the contractor and insists on inappropriate allocation of project risks, then contractors will be motivated to use risk management in their own interest rather than in the interests of the customer. A simple example is claims-seeking behavior by contractors who think they are going to lose money on an onerous fixed-price contract. Enlightened customers expecting risk management from contractors will take pains to demonstrate how contractors can benefit directly by improved cost estimation, greater efficiency, improved project control, and, ultimately, higher profitability.

Similar motivation issues arise in persuading different units in the same organization to undertake risk management. A risk management process that is treated by organization units as just more bureaucracy is unlikely to be fully effective in bringing about proactive management of project risks. In relation to all key players, the risk management process must be seen as immediately useful and valuable, in the sense that it more than justifies the demands made upon them. Furthermore, if the risk management process threatens any of the players, there must be a balance of power in favor of meeting that threat rather than avoiding it.

Summary

Like Simister's, this chapter is concerned with general principles and processes in project risk management that are applicable in all project contexts. These principles can be applied either to *de novo* applications of risk management in particular projects, or, more usefully, to the design of corporate-based formal processes. In the latter case, process design is "strategic" in the sense that an overview of what is appropriate for a given organization is the first consideration. In addition to taking on board the principles discussed under the questions *who?*, *why?*, and *what?*, developing corporate guidelines for deployment of risk management will also involve consideration of *when?* issues in terms of the range of projects that will be subject to risk management processes. In general, comprehensive risk management will tend to be most useful when projects involve one or more of the following:

- Significant novelty (technological, geographical, environmental, or organizational)
- Significant complexity (in terms of technology, organizational structure, political issues, etc.)
- Significant size (in terms of cost, total resources, committed in-house resources, etc.)
- High cost of failure (financially or in terms of reputations)
- Long planning horizons

In time, organizations institutionalizing project risk management may apply different guidelines in terms of the six W questions to different projects, depending on the extent to which the preceding factors are present. However, such sophistication needs to wait on the development of experience with comprehensive risk management processes on selected projects. Failing to consider this issue would be rather like operating a car hire firm that always offers a Rolls-Royce, or a Mini, regardless of the potential customer's wallet or needs. It is difficult to over-emphasize this point because the systematic nature of risk management processes can easily seduce those who ought to know better into the adoption of a single analytical approach for all projects. "If the only tool in your toolbox is a hammer, every problem looks like a nail" is a situation to be avoided.

Even with such sophistication, there will still be considerable need to modify the scope and approach of analysis for individual projects. In effect, the risk management aspect of a parent project needs to be regarded as a project in its own right, requiring a specific design phase prior to execution of risk analysis, and periodic, ongoing development of the process as necessary. This design phase needs to address the six W questions: *who?*, *why?*, *what?*, *whichway?*, *wherewithal?*, and *when?* Addressing the first three of these six W questions refines the scope of analysis and subsequent management actions within the framework of any corporate guidelines. Addressing the latter three questions leads to more detailed, project-specific "tactical" planning of the analysis to be undertaken. However, there are significant interdependencies between these six Ws. For example, in a particular context, *why?* analysis being undertaken may be closely related to who wants the analysis and the scope of risks being included in the analysis. Limiting time and resources may constrain *whichway* the analysis is carried out. Recognition of these interdependencies is important in determining the most appropriate approach to risk management for a given project and organizational context.

References

Ackermann, F., C. Eden, and T. Williams, T. 1997. Modelling for litigation: Mixing qualitative and quantitative approaches. *Interfaces* 27:48–65.

Ashley, D. B., and I. Avots. 1984. Influence diagramming for analysis of project risk. *Project Management Journal* 15(1):56–62.

Chapman, C. B., and S. C. Ward. 2003. *Project risk management: Processes, techniques and insights*, 2nd ed. Chichester, UK: Wiley.

———. 2002. *Managing project risk and uncertainty: A constructively simple approach to decision making*. Chichester, UK: Wiley.

———. 2000. Estimation and evaluation of uncertainty: A minimalist, first pass approach. *International Journal of Project Management* 18:369–383.

Cooper, K. G. 1980. Naval ship production: A claim settled and a framework built. *Interfaces* 10(6): 20–36.

Drummond, H. 1999. Are we any closer to the end? Escalation and the case of Taurus. *International Journal of Project Management*, 17(1):11–16.

Eden, C., T. Williams, F. Ackermann, and S. Howick. 2000. The role of feedback dynamics in disruption and delay on the nature of disruption and delay (D&D) in major projects. *Journal of the Operational Research Society* 51, 291–300.

Goldratt, E. M. 1997. *Critical chain*. Great Barrington, MA: North River Press.

Hartman F. and P. Snelgrove. 1996. Risk allocation in lump sum contracts: Concept of latent dispute. *Journal of Construction Engineering and Management* (September): 291–296.

Hartman F., P. Snelgrove, and R. Ashrafi. 1997. Effective wording to improve risk allocation in lump sum contracts. *Journal of Construction Engineering and Management.* (December): 379–387.

Hillson, David A. 1997. Towards a risk maturity model. *The International Journal of Project and Business Risk Management.* (Spring): 35–45.

Howick, S., and C. Eden. 2001. The impact of disruption and delay when compressing large projects: Going for incentives? *Journal of the Operational Research Society* 52:26–34.

Kelly, J., and S. Male. 1993, *Value management in design and construction: The economic management of projects*. London: Spon.

Klein, J. H. 1993. Modelling risk trade-off, *Journal of the Operational Research Society*. 44(5):445–460.

Markowitz, H. 1959. *Portfolio selection: Efficient diversification of investments*. New York: Wiley.

McCray, G. E., R. L. Purvis, and C. G. McCray. 2002. Project management under uncertainty: The impact of heuristics and biases. *Project Management Journal* 33(1):49–57.

Office of Government Commerce. 2002. *Management of risk: Guidance for practitioners*. London: The Stationery Office.

Uher, T. E., and A. R. Toakley. 1999. Risk management in the conceptual phase of a project. *International Journal of Project Management* 17(3):161–169.

Ward, S. C. 1999a. Assessing and managing important risks. *International Journal of Project Management* 17(6):331–336.

———. 1999b. Requirements for an effective project risk management process. *Project Management Journal* 30(3):37–43.

Williams, T. M. 2000. Systemic project risk management. *International Journal of Risk Assessment and Management* 1:149–159.

Williams, T., C. Eden., F. Ackermann, and A. Tait. 1995. The effects of design changes and delays on project costs. *Journal of the Operational Research Society* 46:809–818.

CHAPTER SEVEN

IMPROVING QUALITY IN PROJECTS AND PROGRAMS

Martina Huemann

In this chapter quality management in projects and programs is described. It begins with a brief history on quality management and an overview on different quality management concepts. The application of general quality concepts such as certification, excellence models, reviews and audits, benchmarking, and accreditation are described in a projects context. In the chapter I differentiate between quality of the contents processes and their outputs and the management processes of the project or program and their outputs. I also make the point that the management quality can be assessed.

When looking in the literature on quality management in projects and programs, we can find a lot for engineering, software, or construction and product development projects. But what quality means in "softer areas"—for example, organizational development—is hardly ever treated. The chapter addresses these projects too, aiming to reflect quality issues on all types of projects.

Project management audit and reviews are described and introduced as learning instruments. Quality management methods—which can be applied to all types of projects, like the project excellence model based on the EFQM Excellence Model—are then described. The chapter concludes with a summary view on quality management and the role of the PM office in the project-oriented company.

From Quality Control to Continuous Quality Improvement

Quality can be defined as the totality of features and characteristics of an entity that bear on its ability to satisfy stated or implied needs (ISO 9000:2000), where an *entity* can be a product, a component, a service, or a process. The need for quality management derives from mass production at the beginning of the twentieth century. Quality management has developed from product-related quality control to company-related Total Quality Management, aiming for continuous process improvement (Seaver, 2003).

Quality Control Based on Statistics

In the early part of the last century, quality control was established in the manufacturing industry to control the quality of parts and products. Quality control became necessary because of the reorganization of the working process introduced by Frederick W. Taylor and the shift to mass production. Regular inspections had to be carried out to find the defective parts and sort them out. When the products became more complex, the work of the inspectors, who were responsible for the quality control, increased. To reduce the quality control costs statistical methods were introduced, based on the assumption that 100 percent quality control is not needed as long as the inspections were done according to statistical parameters. This quality control method is called *sampling inspections*. The objective of quality control was to fulfill an "average outgoing quality level." At that point in time no preventive actions were taken.

During World War II, statistical methods were widely used for mass production in the defense industry in the United States and in Great Britain. Walter Shewhart recognized the need for process control and introduced the first control charts to visualize the variations of the output in a graphical way. Control charts are still used in quality management to observe whether the variations observed are normal process variations or if the process is getting out of control.

Six Sigma, which has gained some attraction in the project management community, is a more recent offshoot of one of the early quality initiatives based on statistics. The Greek letter sigma, σ, is the symbol for standard deviation. It is a measure of variance. The goal of Six Sigma is to reduce process output variation so there is no more than $+/-$ six standard deviations (Six Sigma) between the mean and the nearest specification limit. When a process is operating at Six Sigma, no more than 3.4 "defects" per million opportunities will be produced (Tennant, 2002; Anbari, 2003).

The Deming Approach

Juran, Feigenbaum, Crosby, and Deming are generally considered as the founders of the quality movement. Juran, together with Deming, made a significant contribution to the Japanese quality revolution, published the *Quality Control Handbook* (Juran, 1950, 1986). Feigenbaum, who worked for General Electric, contributed a publication on "Total Quality Control" that already included a lot of features of Total Quality Management (Feigenbaum, 1991). Crosby defined quality as "conformance to requirements" and proved that an organization can get quality for "free" (Crosby, 1979). A detailed presentation of all their approaches is not possible in this chapter. To set the ground, I will concentrate on the approach of Deming (1992).

Deming showed that quality and productivity do not contradict each other but correlate positively, if production is perceived as horizontal process in which the customer relation and feedback for improvement are important features. He describes this as a *chain reaction*. Improvement in quality leads to lower production costs because of less rework, fewer defects, and increased usage of machinery and material. The lower production costs lead to improvement of productivity and make better quality for lower costs possible, which results in an extended market share. This extended market share protects the continued existence of the factory and maintains job.

Later Deming introduced the Deming Circle with the focus on defect correction as well as defect prevention. The Deming Circle, shown in Figure 7.1, consists of the steps "plan, do, check, and act" and is therefore also referred to as the PDCA cycle.

The cycle establishes the base for continuous improvement of the (production) process. If, based on data quality, deficits are detected, a change in the process is planned and tried out on a small scale. The results of this change are checked. If the data improves, the change in the process is introduced. If the improvement did not happen, the cycle is started anew with fresh planning.

Japanese Approach to Quality

In Japan the ideas of Deming and Juran were pursued and further developed by Ishikawa and Taguchi. Ishikawa postulated that all company units and all staff members are responsible for quality and that quality is defined by the customer. He used the word customer in

FIGURE 7.1. PDCA CYCLE BY DEMING.

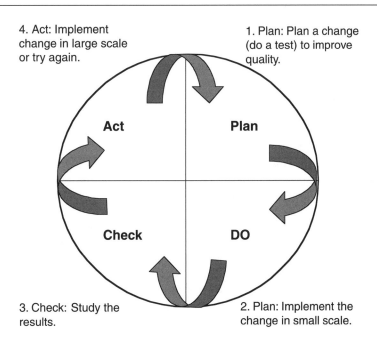

4. Act: Implement change in large scale or try again.

1. Plan: Plan a change (do a test) to improve quality.

Act

Plan

Check

DO

3. Check: Study the results.

2. Plan: Implement the change in small scale.

a broader sense. Customer does not only mean the end customer who pays for the product, but always the next person in the process. Each staff member is both a customer and a supplier at the same time. Further, Ishikawa advocated the statistical methods and introduced the cause-and-effect diagram, also known as a fishbone diagram because of its appearance, which separates the process logically into branches. It is used to visualize the flow of work to determine the cause and effect of problems that are encountered. Quality control circles were established to analyze and to solve quality problems based on Deming's PDCA Cycle (Ishikawa and Lu, 1985).

Taguchi stated that quality is an issue for the whole organization, where quality means conformance to requirements. Statistical process control has been a feature of quality assurance methods for many years, but the application of statistical methods to the design process was his invention. He introduced the concept of "robustness" in design; by that he means the ability of a design to tolerate deviations without its performance being affected (Logothetis and Wynn, 1989; Roy, 1990).

The Japanese word for continuous improvement is Kaizen, a quality management technique that aims for constantly looking for opportunities to improve the process. This has become one of the main features of Total Quality Management. Even if the processes are operating without problems, they are objects for quality improvement. Thus, the quality of a process improves in small increments on a continuous basis (Imai, 1986).

Total Quality Management and Continuous Quality Improvement

Total Quality Management, which was founded by Deming and further developed by the Japanese quality movements, can be summarized in seven principles, which are interrelated (Bounds, Dobbins, and Fowler, 1995).

1. *Customer orientation.* The customer buys a product or service because of the benefits to be gained from it. Therefore, the needs and requirements of the customer have to be known. But "the customer" is not only the external customer; the internal customer in the process is also relevant. Customer orientation is the main principle of Total Quality Management.
2. *Continuous improvement of systems and processes.* This requires also that the quality standards and procedures have to be continuously checked and further developed.
3. *Process management.* Performance is based in the competencies of the staff members to fulfill the business processes in the company. Most of the defects have their origin in inadequate processes. Blame the system instead of blaming the worker. Table 7.1 compares the attitude of traditional managers with process managers (Harrington, 1991, p. 5).
4. *Search for the true reason.* The basic assumption is that any problem is only a symptom and has a true origin in the system. Only by digging into the process can the true reason be found and a repetition of the mistake be prevented.
5. *Data collection and analysis.* Data has to be collected as a basis for improvement. For quantitative data collection and analysis, statistical methods such as process control carts, cause-and-effective diagrams, and Pareto charts can be used. Additional, more qualitative methods are audits, reviews, and benchmarking.

TABLE 7.1. TRADITIONAL MANAGERS IN COMPARISON WITH PROCESS MANAGERS.

Traditional Managers	Process Managers
• The staff member is the problem.	• The process is the problem.
• Do the own job.	• Support the others to do the job.
• Know own job.	• Know own job within the whole process and context.
• Evaluate staff members.	• Evaluate process.
• Change staff members.	• Change process.
• Search for a better staff member.	• The process can always be improved.
• Control staff members.	• Train and further develop staff members.
• Find the guilty one to blame for a mistake.	• Find the reason for the mistake.
• The quantitative result is important.	• Customer orientation is important.

6. *People-orientation.* A continuous improvement is only possible if the competencies and skills of the staff members are also continuously further developed. The employees have to be educated, trained, and empowered. In high-performance organizations, people are *enabled* to do their best work. They have the adequate tools, standards, policies, and procedures.

7. *Team-orientation.* Team orientation is part of the culture and structure of the organization. Members of one team also communicate and cooperate with members of other teams to fulfill a process. That makes the traditional department-oriented thinking obsolete and changes it into a process-oriented, trans-sectional way of thinking.

Overview on Quality Management Methods in the Project-Oriented Company

In project-oriented companies that apply modern quality management based on Total Quality Management and Continuous Improvement, different quality management methods—often in combination—are applied. These include the following:

- Certification
- Accreditation
- Excellence model
- Benchmarking
- Audit and review
- Evaluation
- Coaching and consulting

Certification

Certification is a procedure in which a neutral third party certifies that a product, process or a service meets the specified standards. An important—but by no means the only—certification process is ISO Certification according to the International Standards for Quality (ISO). The ISO is a network of national standards institutes from 147 countries working in partnership with international organizations, governments, industry, business, and consumer representatives. ISO has developed over 13,000 International Standards on a variety of subjects (www.iso.ch/).

One advantage of ISO certification is that it is international. For example, the ISO 9000:2000 series of standards are far more process-based than ISO 9000:1994, which were mainly based on procedures. This was a shortcoming, as ISO auditors ended up ticking boxes whether a document was there or not, without considering any further quality dimensions. The ISO 9000:1994 became obsolete by December 2003. The new ISO 9000:2000 also considers the quality of results and is therefore compatible with excellence models and accreditation schemes. Therefore, the new series will fit better to the process philosophy of modern quality management in project-oriented companies.

For projects and programs, the new ISO 10006:2003, "Quality Management Systems: Guidelines for Quality Management in Projects," exists, which provides a structured approach for the optimal management of all processes involved in the development of any type of project.

Other certifications relevant in the context of project management and project-oriented organizations are certifications done on the PRINCE2 project and MSP program management methodology (OGC, 2002) and professional certification by IPMA, IPMA member organizations, and PMI for individual project management personnel.

Accreditation

An accreditation is an external evaluation based on defined and public known standards. Accreditation was originally established to support customer protection. Consumers can be protected by certification, inspection, and testing of products and by manufacturing under certified quality systems. Consumers need confidence in the certification, inspection, and testing work carried out on their behalf, but that they cannot check for themselves. The certifiers of systems and products as well as testing and calibration laboratories need to demonstrate their competence. They do this by being accredited by a nationally recognized accreditation body. Accreditation delivers confidence in certificates and reports by implementing widely accepted criteria set by, for instance, the European Committee for Standardization (CEN) or international (ISO) standardization bodies. The standards address issues such as impartiality, competence, and reliability; leading to confidence in the comparability of certificates and reports across national borders. (See www.european-accreditation.org/).

Accreditation is a commonly used quality management method, for instance, in the healthcare sector, in which a lot of projects and programs are carried out. There the accreditation was established to protect staff members and patients from faulty organizational processes. Participating in an accreditation program is voluntary. The applying organization does a standardized self-assessment. The results of the self-assessment are the basis for a site visit, where the surveyor uses documentation analysis, observations, and interviews for information gathering. Results of the site visit are summarized in a report. The applicant gives feedback to the report. This feedback discussion can be organized in the form of a workshop. Then the final result of the accreditation—which can be numerical or a descriptive like "substantial compliance, partial compliance, minimal compliance, or noncompliance"—is provided. Accreditations have to be renewed every couple of years.

In the project management context, accreditation is done, for example, by PMI for project management education and training programs. The degree and nondegree programs are accredited for their content and progress compliance with the standards set by the Global Accreditation Center for Project Management. (see www.pmi.org)

Excellence Model

Excellence models are nonnormative models that provide a framework to assess an organization in it's degree of excellence in the application of practices. All excellence models differentiate between enabler criteria and result criteria as a basis for the assessment. The most important excellence models have been developed in the frame of regional quality programs, which award organizations for outstanding quality improvement. I mention here the following:

- Deming Prize
- Malcom Baldridge National Quality Award
- European Quality Award
- International Project Management Award

Deming Prize. The Union of Japanese Scientists and Engineers (JUSE) created a prize to commemorate Deming's contribution to quality management and to promote the development of quality management in Japan. The prize was established in 1950 and annual awards are still given each year (See www.deming.org/demingprize/)

Malcom Baldridge National Quality Award. The Baldridge Award is given by the President of the United States to businesses and to education and healthcare organizations that apply and are judged to be outstanding in seven areas: leadership, strategic planning, customer and market focus, information and analysis, human resource focus, process management, and business results. Congress established the award program in 1987 to recognize U.S. organizations for their achievements in quality (see www.quality.nist.gov)

European Quality Award. The European Foundation for Quality Management (EFQM) was founded in 1988 with the endorsement of the European Commission. It is the European framework for quality improvement along the lines of the Malcolm Baldridge Model in the United States and the Deming Prize in Japan. The European Model for Business Excellence—now called the EFQM Excellence Model—was introduced in 1991 as the framework for organizational self-assessment and as the basis for judging entrants to the European Quality Award, which was awarded for the first time in 1992 (see www.efqm.org/).

International Project Management Award. An excellence model in the context of projects is the project excellence model, which is based on the European Model for Business Excellence. The project excellence model is described in detail in a later section in this chapter (see www.gpm-ipma.de).

Benchmarking

By the end of the 1970s companies like Xerox in the United States were suffering from competition from Japan. They had to find out what their competitors were doing. Benchmarking as a tool to compare the performance and practices of one company with other companies derived from the work of Robert Camp (1989). The aim is to understand the reason for the differences in performance by examining the process in question in detail. Benchmarking is a tool for improving performance by learning from best practices and understanding the processes by which they are achieved. Application of benchmarking involves following basic steps:

1. First, understand in detail your own processes.
2. Next, analyze the processes of others.
3. Then compare your own performance with that of others analyzed. Comparison can be done within one's own organization or with other organizations from the same industry or different industries.
4. Finally, implement steps necessary to close the performance gap.

A number of benchmarking models and processes have been developed and are applied for a wide range of subject areas. In the context of project-oriented companies, maturity models are often used as a basis for the benchmarking exercise. Most of the models are used for benchmarking on the company level and not on the single-project level. Also, specific benchmarking communities for project and project management benchmarking exist. Most of these benchmarking activities are industry-specific.

Audit and Review

ISO 19011:2002 defines auditing as a "systematic, independent and documented process for obtaining audit evidence and evaluating it objectively to determine the extent to which the audit criteria are fulfilled." The audit criteria are a set of policies, procedures, or requirements. Reasons for audits are, for instance, certification, internal audits and review, and contract compliance. Reviews are considered to be less formal than audits.

Audits and reviews are applied to ensure quality in projects and programs. Peer reviews for projects and programs are done by peer professionals such as program managers, project managers, or other experts who are not part of the project or program under consideration. Audits and reviews are not meant to be a replacement for other exchange of experience activities, like coaching or experience exchange workshops. Audits and reviews as methods of quality assurance and quality improvement in the context of projects and programs are further described later in this chapter.

Evaluation

In general, evaluation is referred to as a systematic inquiry of the worth or merit of an object. Evaluations are applied for projects and programs. While audits and reviews are performed during the project or program, evaluations are carried out when the project or program is finished. Objects of evaluation are the management processes, the technical processes, and performance criteria.

Coaching and Consulting

Coaching and consulting might not be considered as part of quality management at first sight. But management consulting on projects and programs as well as management coaching of project and program managers are definitely quality management methods to ensure management quality. These methods are applied in many advanced project-oriented companies. While in management coaching the client is an individual—for example, the project manager or program manager—the object of consideration in consulting is the project or program. Coaching is often considered as a method to further develop personnel. If a new project management approach has been implemented in a company, coaching may be provided to project managers to support the implementation. In this way the quality of the management process is ensured.

A typical situation for management consulting is a program start-up. The situation is typically rather complex but very important for the success of the program, as in the start-up process, the quality for the management of the program is set. Another typical situation for management consulting is a project or program crisis. Then the consultant helps to manage the discontinuity. Consulting activities can also support the program to implement the corrective and preventive actions, which have been agreed on after a management audit of the program. Large project-oriented companies often provide these services through their PM office and have internal PM coaches and management consultants for projects and programs.

Application of Quality Management Methods in the Project-Oriented Company

All quality methods described in the preceding text are applied in the project-oriented company. They are often combined based on a continuous-improvement philosophy. Some of the methods described are commonly applied, such as certification, benchmarking, and design audits, while others are rather new, such as PM audits, PM coaching, and PM consulting. In general, for project-oriented companies quality management is relevant for the following:

- The project-oriented company as such
- The company units (e.g., human resources department, engineering department, PM office, finance department)
- The single program
- The single project

In this chapter I concentrate on the quality management issue of projects and programs and look at some of the quality management methods in more detail. Still, the quality management of the project-oriented company as such is an important context to the quality management in projects and programs, as the company should have a quality system in place that provides quality standards and procedures for the projects and programs.

Quality Standards for Projects and Programs

Product and process quality in projects and programs

Good quality in the context of projects and programs is defined (Turner, 2000) as being to

- meet the customer requirement
- meet the specifications
- solve the problem
- fit the purpose
- satisfy or delight the customer.

Possible objects of considerations regarding quality in projects and programs are as follows:

- The product (services)—that is, *the material results* of the project or program, which could range from a marketing concept, a feasibility study, a conference, a new product line, or a power plant
- The project or program *content processes* to create the project or program results (e.g., designing, engineering, implementing, and testing)
- The *management processes* in the project and program: These include project management processes such as project start-up, project coordination, project controlling, and project closedown.

For each of these objects of consideration, standards need to be provided in the project-oriented company to set the basis for quality management. We can think of this in knowledge and learning terms. Organizations have the capability to gather knowledge and experience and to store such knowledge in a "collective mind" (Senge, 1994). The organizational knowledge is hidden in the systems of organizational principles, which are often anonymous and autonomous; these define the way the organization works. Through these, one can discover the organization's knowledge and experience, operation procedures, description of work processes, role descriptions, recipes, routines, and so on.

Standards as Basis for Quality in Projects and Programs

In general, standards can be

- generic—therefore applicable to all types of projects and programs in all project-oriented companies of any industry; or
- specific—includes specifics for a certain industry, company, project, product, or customer.

Further, standards can be normative or non-normative. There are different types of standards relevant for projects and programs, which are

- product standards
- project and program standards
- project management standards
- other standards, such as procurement standards, health and safety regulations, and so on.

Product Standards. Product standards can either be based on regulations or common shared standards or specific agreements with the customer. These product specifications covers issues like the following:

- The required functionality of the product or service
- Design standards
- Cost of the product or service and time at which is should be delivered
- Availability, reliability, maintainability, and adaptability

Project specifications are agreed on with the customer and (internal) project sponsors. Some companies have company-internal specific product quality and safety regulations, which then also apply to their product development projects. In some industries, like the pharmaceutical industry, there are very strict regulations by authorities, which of course then also apply to project work. For repetitive projects, standardized product breakdown structures may be applied.

Project and Program Standards. Project standards are normally project-type-specific and provide a standard procedure for the content and management processes of the project or program, as in the following examples:

- Phases for a construction project: concept, definition, design, supply, construction, commissioning
- Phases for a software development project: Requirement definition, requirement specification, prototyping, design specification, system development, test system, install system, pilot application

Project-oriented companies may wish to further standardize repetitive projects by providing detailed project process descriptions. Project work breakdown structures and even work packages specifications may be standardized. This is very often the case in industries with a lot of repetitive projects such as in the engineering industry.

Figures 7.2 and 7.3 show an example for standardized processes and their relation to project standards. The company is part of an international concern and acts as supplier for standardized intermediates and high-quality fine chemicals for the life science industry (pharma, agro, food). In this project-oriented company, different processes have been defined, of which parts are carried out in the form of projects. Different types of projects are carried out, of which R&D projects as well as investment projects have a repetitive character (Stummer and Huemann, 2002).

For all repetitive projects, the company provides standard project plans, with the following objectives:

- To ensure the quality of processes and results
- Therefore to optimize project and program costs and duration
- To enable the use of existing knowledge instead of permanently reinventing the wheel

FIGURE 7.2. TYPES OF PROJECTS IN A CHEMICAL COMPANY.

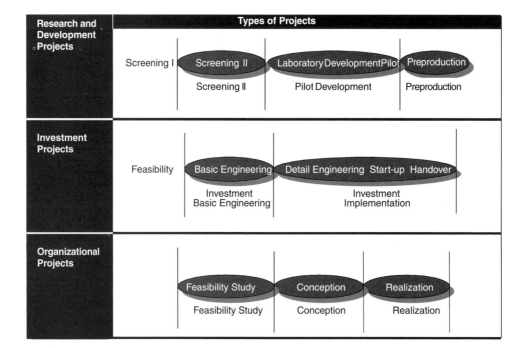

FIGURE 7.3. EXAMPLE: STANDARD PROJECT PLAN:
WBS LABORATORY DEVELOPMENT.

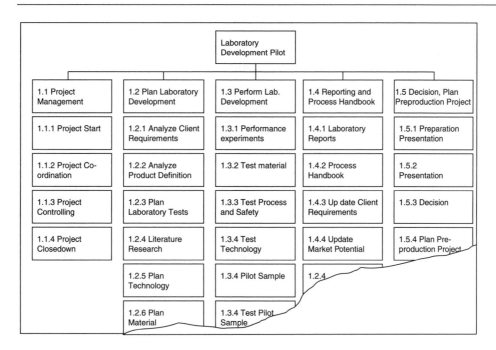

In this company there are standard project plans such as project objectives plan, list of objects of consideration, work breakdown structures, work package specifications, project milestone list, project organization, and so on. The standard work breakdown structure of a laboratory development project is shown in Figure 7.3.

Project Management Standards. Generic PM standards are standards that are applicable to all types of projects; examples for generic project management standards are PRINCE2 (OGC, 2002) the PMI's PMBOK (PMI, 2002), and pm baseline (Gareis, 2002). For a detailed discussion, see the chapter by Crawford.

Many project-oriented companies have implemented their own project management guidelines, often based on one of the international project management standards. Table 7.2 shows a content structure of guidelines for the management of projects and programs based on the pm baseline. These PM guidelines limit the variety of different management approaches and methods to a common company standard and provide means of support and help to ensure the program and project management quality in the single program or project. The guidelines prevent every project manager having to invent the same processes again and again.

TABLE 7.2. GUIDELINES FOR THE MANAGEMENT OF PROJECTS AND PROGRAMS.

Guidelines for the Management of Projects and Programs

Checklists and standard project handbooks may also provide the structure for the project manager and his or her team to manage the project.

Quality Management in Projects and Programs

PMI's PMBOK (PMI, 2002) states that project quality management includes the processes required to ensure that the project will satisfy the needs for which it was undertaken. Hence, says PMBOK, quality management includes the following:

- *Quality planning.* To identify all the quality standards relevant for the project and plan how to satisfy them
- *Quality assurance.* To evaluate the project to ensure that the relevant quality standards will be met
- *Quality control.* To monitor, to compare with the relevant quality standards, and to correct the product (components, their configuration, the facility) and the processes

Traditional quality management approaches for projects and programs, says PMBOK, concentrate more on the product quality as such and on quality control by statistical means like inspections, control charts, Pareto diagrams, statistical sampling, and so on. This perception is rather shortsighted, as projects need more than manufacturing quality management approaches. Quality assurance by using reviews and audits becomes more important in projects. The processes need to be checked rather early to ensure the quality of the project deliverables, as only sound processes lead to good products and solutions.

Quality of Processes as Basis for Product Quality

An example for quality planning and assurance in a program is the Austrian Railway revitalization initiatives. Within this program 40 railway stations were revitalized. The program work breakdown structure is shown in Figure 7.4 (below). To plan the quality management of the program, a standardization project was performed to develop process standards, to introduce a documentation management system and to collect all relevant standards to be applied within the program. Within the program, three types of repetitive projects existed, namely, conception projects, planning projects, and realization projects. For these repetitive projects, standard project plans including work package specifications were developed. The project outputs were also standardized. For instance, for each of the railway stations, a feasibility study was carried out. The structure of how this study should look was standardized, so the feasibility studies were comparable.

Quality Management in Engineering, Construction, and IT/IS Projects and Programs

Solution quality and process quality in engineering, construction, and IT/IS projects and programs are typically assured as follows (Turner, 2000):

FIGURE 7.4. WORK BREAKDOWN STRUCTURE OF AN INFRASTRUCTURE PROGRAM.

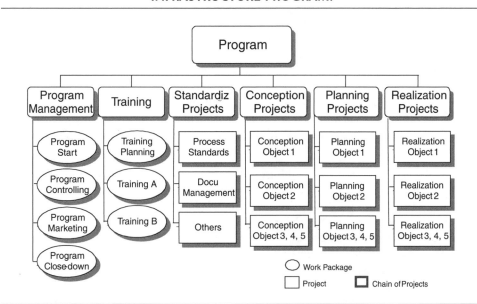

- *Competent project personnel.* Previous experience of the project team and the project-oriented organization in creating the facility
- *Well-defined specifications.* These include required functionality of the facility and its components, design standards it is required to meet, the time and cost at which it should be delivered, availability, reliability, maintainability, and adaptability.
- *Standards.* These include standards the solution has to meet and process standards the project should follow.
- *Audits and reviews of the project deliverables.* These include, for example, design reviews (Webb, 2000) of the project processes.
- *Change control and configuration management.* As described in the chapters by Kidd and Burgess, and by Cooper and Sklar Reichelt.

Audits and reviews to check the project processes and results are the most common methods for quality assurance in projects and programs. They are applied at the end of phases. For example, the gate model for process improvement of an international engineering company—applicable for their integrated solution delivery programs—shows the phases: concept, design, development, implementation, and benefits delivery. Reviews are carried out to evaluate the deliverables produced during the phases. The reviews are assessments of the solution under development, which include the following:

- *Concept phase review(s).* To assess the completeness of the design concepts, including consideration of alternative designs.
- *Design phase review(s).* To assess the completeness of the design phase work, which include, for instance, process design and system requirements, logical design, operations plan, and test plan.
- *Detailed design review.* To do a complete technical assessment of the detail design before beginning extensive coding or purchasing of software.
- *Pilot readiness review.* To assess whether the solution is ready to pilot.
- *Implementation readiness review.* To assess the readiness of implementing the solution to its planned full extent.
- *Implementation reviews.* To assess the implementation on each site that implements the new solution. It includes validation of implementation measurements, system performance, site adjustments, planning adjustments, implementation logistics, budget, and schedule.

The reviews are linked to the gates, which are go/no-go decision points. Only successful reviews allow a project to schedule a gate meeting. Beyond these reviews, they are tested by third parties to ensure that the solution is in accordance to specifications. In concurrent engineering the gates are difficult to specify, leading to Cooper's notion of fuzzy gates and other flexible project checks.

These quality assurance activities are an inherent part of the technical content processes and have to be visualized in the work breakdown structure, the Gantt chart, and the cost plan of such a project. To have such reviews and to improve the solution on the way is part of best-practice quality management for these kinds of projects. But what about other types of projects and programs?

Quality Management is an Issue in All Types of Projects and Programs

The approaches to quality management in projects and programs typically quoted reflect those types of projects that result in a facility or system, such as a power plant, a new product, or an IT system. While some of the quality management issues mentioned previously, like competent project personnel, are relevant for other projects like organizational development, event management, marketing, or indeed any kind of project, quality issues like design reviews or product standards or configuration management may not be. Either they are not applied at all or different terminology is used.

In organizational development projects and programs, the best solution is no good if it is not accepted by the organization. So quality is very much linked to acceptance. This means that the process has to be designed so that people can get involved and contribute to the solution (Frieβ, 1999). Instead of reviews to ensure the quality of the solution in organizational development projects, working forums like workshops or presentations are often more relevant. In workshops, solutions are commonly created and evaluated. Cyclic working, doing first a draft then presenting it, getting feedback, and including the feedback improves the quality. Typically, no standard program plans for such programs are available in the organization, as they rarely have repetitive character. Like the railway station program shown earlier, program standards can be developed within the program to ensure the quality of the content's processes and of the deliverables. Project and program management is responsible for the adequate design and the management of the content's processes. Thus, high-quality program and project management is crucial for these kind of projects and programs of high social complexity.

Management Quality of Projects and Programs

The quality of the deliverable that is created in the project—whether it is a marketing concept, a new organization for a company, a newly implemented IT system, or an international congress for 500 PM experts—depends always on the processes of creating it. If there are no standard project processes, it is the task of project management to design these content processes and the adequate organization by providing a process-oriented work breakdown structure, appropriate schedules, adequate organizational roles and responsibilities, and so on. It is also a project management task to manage these processes by controlling the status, adapting plans if necessary, agreeing on new responsibilities, and so forth.

The less repetitive a project, the more parties are involved in the project; therefore, the more socially complex, the more important is the management process and its quality. The quality of the project management process can be measured by looking at the design of the management process and at its results (Gareis and Huemann, 2003).

- Design elements of the management process are, for example, participants taking part in the management, working forms applied, infrastructure used—there is no question that there is a difference in quality of the project plans if they are only prepared by the project manager behind his or her computer or if they are the result of a two-day start-up workshop for the project team, including customer and supplier representatives.
- Results of the start-up process for a project or program will include, for example, adequate program plans like a work breakdown structure, objective plan, schedules, cost

plans, and so forth; adequate consideration of the program context; adequate program organization and communication structures; agreed on responsibilities; an established program management team; an established program office; an appropriate program culture; and so on.

The project organization contributes to the quality of the solution delivered. High-quality results can be promoted by building an integrated project organization with suppliers and customers working in one project organization together instead of building parallel investor, contractors', and subcontractors' project organizations. If this is based on a common contract, it is called *partnering* (Scott, 2001). This form of project organization saves time, reduces costs, and improves quality.

The principle of partnering is rather simple: Performance is improved by buyers and suppliers working together closely on a long-term Total Quality base. Aims are matched, resources pooled, teamwork promoted, and planning integrated (Morris, 1999). One documented case is the Ruhr Oel case (van Wieren, 2002), an alliance of Ruhr Oel, the client; Fluor Daniel, the engineering contractor; and Strabag, Fabricon, Ponticelli, the construction contractors. All partners had their own quality management system; the Fluor Daniel system formed its basis. They had common project execution plans and alignment meetings to share information. An alliance board, which consisted of senior managers of each company and the project managers from Ruhr Oel and Fluor Daniel met monthly to monitor the project and check that all parties remained aligned. The best-qualified person was nominated for each key position. Strong cooperation between and integration of all team members from all companies was supported. As no claims were possible according to their alliance contract, the only way to solve conflicts was through communication. The alliance managed to complete the EPC work for two refineries in one year and seven months. Although despite a 25 percent increase in work due to scope development during the EPC phase, the project was completed on time and 9 percent under the target price. The plants were operating as specified and meeting quality standards, and all authority requirements were fulfilled. By the way, the project won the International Project Management Award in 2002.

Management Audits and Reviews of Projects and Programs

A project audit is a systematic and independent investigation to check if the project is performing correctly with respect to project and or project management standards. A *project review* is defined as a formal examination of the project by persons with authority in order to see whether improvement or correction is needed. Another word for project review is project health check (Wateridge, 2000). A special form of the project review is the peer review: Here the review is carried out by experienced peer project managers to give feedback and advice to the project.

In the previous section the review of the deliverables was introduced as one method of quality assurance commonly applied in construction, engineering, and product development projects and programs. In this section I concentrate on management audits and reviews of projects and programs that are applied in all kinds of projects and programs. Management audit and reviews of projects and programs assess the management competencies of the projects or program, namely, the organizational, team, and individual competencies to perform the management processes. Thus, the project or program management process and its results are reviewed. Results of the project start process could be, for instance, that adequate project plans exist and the project team has been established. The benefits of management audits and reviews of projects are on the one hand to provide a learning opportunity to the single project to improve its project management quality. On the other hand, by evaluating the results of several management audits, patterns can be found. For instance, if a lot of the projects have a low-quality cost plan or do not apply a stakeholder analysis, this shows that these issues are general subjects for improvement in the project-oriented company (Huemann and Hayes, 2003).

Differences between Management Audit and Review

Table 7.3 gives an overview of the differences between management audits and reviews. The main differences can be seen in terms of obligation and formalization. We consider reviews to be less formal and the obligation to implement the review results is medium in comparison with audits. Management reviews serve the purpose of learning, feedback, and quality assurance, while management audits are also used for problem identification and controlling.

We differentiate between the initiator and the owner of the management audit or review. The management audit or review owner is the project or program owner, as he or she has to provide the resources. The audit can also be initiated by a profit center, the PM office, or the customer. The management review is often initiated by the project manager or program manager. The client is always the project or the program and not the single project manager or program manager. Management audits and reviews are always carried out by persons who are external to the project. The auditors or reviewers are never part of the project or program organization.

PM Audit Criteria Depend on the PM Approach Used

The PM audit criteria depend on the PM approach used. The basis for the PM audit is the PM procedures and standards. If the PM audit is based on a traditional PM approach, the audit criteria are limited to the PM methods regarding scope, schedule, and costs. Additional PM objects of consideration like the project organization, the project culture, and the project context become only PM audit criteria, as, for example, if PRINCE2 is used. If project management is considered as a business process consisting of the subprocesses' project start, project controlling, project coordination, management of project discontinuities, and project closedown, the design of the PM process becomes an audit criterion. The PM approach used in a PM audit has to be agreed on.

TABLE 7.3. DIFFERENCES AND COMMONALTIES OF MANAGEMENT AUDIT AND REVIEW.

	Management Audit	Management Review
Initiator	• Project or program owner • Profit center • PM office • Customer	• Project manager or • Program manager
Owner	• Project owner or • Program owner	• Project owner or • Program owner
Client	• Project or • Program	• Project or • Program
Purpose	• Learning • Feedback • Controlling • Problem identification • Quality assurance	• Learning • Feedback • Quality assurance
Obligation	• High	• Medium
Formalization	• High	• Medium
Methods	• To be agreed on • All possible	• To be agreed on • All possible
Object of consideration	• Management process(es) and results • Organizational, team, and individual project or program management competence	• Management process(es) and results • Organizational, team, and individual project or program management competence
Homebase of auditors / reviewers	• Company external • Company internal	• Company internal (peer review)
Number of auditors/reviewers	• 1–3	• 1–3
Duration	• 1–2 weeks	• 2 days–1 weeks
Resources	• Depending on scope of the project or program and methods agreed on: 8–12 days	• Depending on scope of the project or program and methods agreed on: 2–8 days

Adequate Times for PM Audits

PM audits can be done randomly, regularly, or because of a specific reason. They are still very often carried out if somebody in the line organization has a bad feeling about the project. Then the method is used for problem identification and controlling, not so much for learning purposes and quality assurance. Nevertheless, the ideal point in time to do a project management audit is in a relatively early phase of the project—for instance, after the project or program start has been accomplished. That gives the project the chance to further develop its management competence. Further PM audits/reviews later in the project are possible to give further feedback but also to verify if the recommendations agreed on in earlier PM audits were taken care of by the project or program.

PM Audit Process

An audit needs a structured and transparent approach (Corbin, Cox, Hamerly and Knight, 2001). A PM audit process established in a project-oriented company in accordance with the ISO (19,011: 2002) includes following steps.

- Situation analysis
- Planning PM audit
- Preparation PM audit
- Performance of analysis
- Generation of PM audit report
- Performance of PM audit presentation
- Termination of PM audit

The follow-up of the PM audit, thus verifying if the corrective and preventive actions recommended by the PM auditors have been implemented in the project, is not part of the PM audit process. There is the need for an agreement between representatives of the project and the project owner as to which of the actions recommended by the PM auditors have to be implemented. A PM audit follow-up agreement form is shown in Figure 7.5.

Methods of PM Audits

In the PM audit, a multimethod approach is used. The following methods can be applied for gathering information:

FIGURE 7.5. PM AUDIT FOLLOW-UP AGREEMENT FORM.

PM Audit Follow-up Agreement		
Name of project audited:	Name of project manager:	
PM audit completed at:	Name of project owner:	
Name of auditors:	Date of follow-up agreement:	
Corrective Actions		
Action	Responsibility	Deadline
Preventive Actions		
Action	Responsibility	Deadline
.... Project Manager		.... Project Owner

- Documentation analysis
- Interview
- Observation
- Self-assessment

For presenting the PM audit findings, the following methods can be used:

- PM audit report
- PM audit presentation
- PM audit workshop

Documentation Analysis. In a documentation analysis, the organizational PM competence of a project can be observed. Documents to be considered are PM documents like, for example, project work breakdown structure, project bar chart, project environmental analysis, project organization chart, project progress reports, and minutes of project meetings. The auditors can check whether or not the required PM documents exist. This might end up in ticking boxes without checking whether the contents of the document make sense. A further step is to audit also the quality of the single PM document, as well as the constancy between the single PM documents. Figure 7.6 shows an example of a checklist for assessing the quality of a work breakdown structure. Criteria are as follows:

- Completeness
- Structure
- Visualization
- Formal criteria

FIGURE 7.6. AN EXAMPLE FOR A PM AUDIT CHECKLIST.

PM Audit: Work Breakdown Structure			
Project:	Company:		
Criteria:	Weight	Result	Weighted Result
Completness			
Structure			
Visualisation			
Formal Criteria			
Total			
PM Auditor:	Date:		Page:

The criteria used for checking the quality very much depend on the project management standards of the company. If they use a project management approach that promotes the use of the PM plans as communication instruments in the project team, then the criteria "visualization" becomes important. The criteria might even be weighted. So, for instance, to fulfill the "formal criteria" is less important than having a complete project plan.

Interview. Interviews are conducted to obtain more detailed information based on questions that arise from the documentation analysis. One can differentiate between group interviews and interviews conducted with a single person. In the case of group interviews, the PM auditor has the opportunity to see how the interview partners interact with each other and react to different opinions. In management audits of projects, interviews can be held with representatives of the project organization, the project manager, the project sponsor, the project team members, as well as with representatives of relevant environments like, for example, the customer, suppliers, and so on. When doing interviews with the customer, which might be quite sensitive, the PM audit result is based on information provided from different angles. That approach can be compared with a 360-degree feedback approach. The customer and suppliers can give another perspective from outside, which might be very different to the perspective the project team has.

Observation. In the observation, the PM auditors collect further information about the project management competence in the project by using observation criteria. Project owner meetings, project team meetings, and project subteam meetings can be observed.

Self-Assessments. Within the PM audit, self-assessments of the individual PM competence of representatives of the project organization—for example, project manager, project owner, and project team member—can be applied (Huemann, 2002). In some project-oriented companies, such assessments are applied on a regular basis as part of the human resource management to further develop the PM personnel. Then the self-assessment of the individual PM competence is not part of the PM audit.

Further, a self-assessment of the PM competence of the project team can be carried out in a PM audit. The PM competence of the project team can be described as the knowledge and the experience of the project team to develop commitment in the project team, to create a common "big project picture," to use the synergies in the project team, to solve conflicts, to learn in the project team, and to jointly design the PM process. These self-assessment activities very much can make the PM audit a learning experience for the project or program team.

PM Audit Reporting. The objective of the PM audit report is to summarize the findings of the PM audit and give recommendations regarding the further development of the PM of the project. It also includes recommendations for the further development of the project management in the project-oriented company. The PM audit report is the basis for the follow-up agreement between the project and the PM audit owner on which actions have to be taken. Table 7.4 shows the structure of a PM Audit report as an example. The PM auditors are not responsible for checking whether their recommendations are followed. How-

TABLE 7.4. STRUCTURE OF THE PM AUDIT REPORT.

1. Executive Summary
2. Situation Analysis, Context, and Description of the PM Audit Process of Project XY
3. Brief Description of the Project XY
4. Analysis of the PM Competence of Project Management of Project XY
4.1 Analysis of the Project Start
4.2 Analysis of the Project Coordination
4.3 Analysis of the Project Controlling
5. Further Development of Project Management of Project XY
6. Further Development of Project Management in General
7. Enclosures

ever, if there is a PM audit at a later point in time, the PM auditor will also have a look at the previous PM audit report.

PM Audit Presentation. The PM audit reporting will always be in written form. Often there is also a PM audit presentation, before the written report is handed in by the auditors. Participants in the PM audit presentation are the PM audit owner, project manager, and further representatives of the project. Further representatives of relevant environments of the projects—for example, representatives of the client, supplier, and so on—can be invited. In many project-oriented companies, the PM audit presentation is considered as important. The objectives of the PM audit presentation, sometimes even organized as workshops, is to understand the PM audit results. That leads to more acceptance of the PM audit results and provides the chance to the project to become a learning organization as defined by Peter Senge in *The Fifth Discipline Fieldbook* (1994).

Application of Adequate PM Audit Methods

Which methods for information gathering are used depend on the specific case and on the agreement between the PM audit owner and the project manager of the project to be audited. The PM audit should at a minimum (ISO 19011) include documentation analysis and interviews at least with the project manager, the project owner, and representatives of the project team. In the case of a program, interviews with the program owner, the project managers, and project sponsors of the different projects are required. In the case of projects, further interviews with representatives of relevant environments like client and suppliers are important. By observing meetings, the PM auditors get an insight in the PM team competence. The self-assessments help achieve a holistic picture. Self-assessments provide the individuals and the project team the possibility to reflect and very much add to the learning perspective of the PM audit. If the individual PM competence of PM personnel is assessed by the project-oriented company, this self-assessment will not be part of the PM audit. The quality of the results of the PM audit depends very much on the scope of the methods and the professional application of these.

PM Audit Organization and Roles

In the PM audit system, the roles PM audit owner, PM auditor, representatives of the project, and representatives of relevant environments can be differentiated. The PM audit owner is responsible for the assignment of the PM audit and for agreeing the scope and timing of the PM audit with representatives of the project. Further, the PM auditor has to ensure resources for the PM audit. Often the PM audit is performed by two to three auditors. Then one of the PM auditors takes over the role of the lead auditor. The PM auditors analyze the PM competence of the project and give recommendations regarding the further development of the project management of the project. The PM auditor needs not only profound PM competencies but also audit competencies like designing the PM audit process or performing an interview professionally. Thus, social competence and emotional intelligence are important.

The role of the representative of the project is taken over by the project manager of the project audited. The objective of this role is to contribute information for the PM audit and to invest resources. Tasks of the project manager in a PM audit are, for example:

- Contribution to clarify the situation in the project
- Feedback to the PM audit plan
- Agreeing scope and methods of the PM audit
- Interview partner in the PM audit
- Provision of PM documents of the project for the documentation analysis.

PM Audit Values and Limits

The communication policy should be agreed on between the PM auditor and project manager at the beginning of the PM audit. From a learning perspective, the PM audit should be done in a cooperative and not a hostile way. This would also mean that the project manager of the project that is audited should be kept informed by the PM auditors. Circumstances that should lead to a cancellation of the PM audit and the consequences of a cancellation should also be agreed on at the start. One major challenge is that the audit result is not perceived as a feedback to the single project manager who then will be blamed for mismanagement. This requires a certain culture of openness in the project-oriented

FIGURE 7.7. PM AUDIT SYSTEM.

company. A good example is CMG, a global consultancy company that has a long tradition in performing management audits and reviews of projects and programs. In 2002 about 7 percent of their annual turnover was spent on quality management. About half of it was spent for management audits of projects and programs, which also include the training of the auditors in doing their job correctly.

The International Project Management Award

Project Excellence Model

The International Project Management Award is based on the project excellence model. The project excellence model was developed by the German Project Management Association (GPM; *www.gpm-ipma.de*) for the IPMA and is based on the EFQM Model. The project excellence model is applicable to any project type. There is no specific consideration of programs. The model, as shown in Figure 7.8, altogether assesses nine criteria divided into two sections: Project Management and Project Results: The Project Management section

FIGURE 7.8. PROJECT EXCELLENCE MODEL.

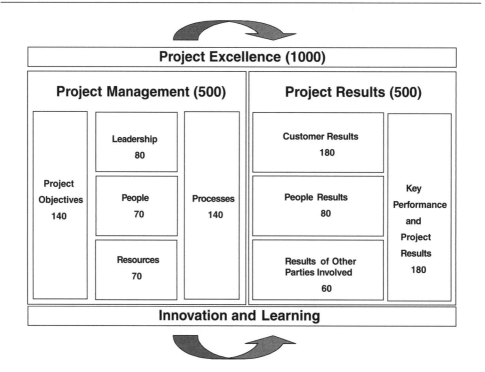

evaluates how far the enabler processes are excellent, while the Project Results section evaluates the degree of excellence of the project results.

The Assessment Criteria

The criteria for the assessment of project management include the following:

- *Project objectives.* How the project formulates, develops, checks, and realizes its objectives
- *Leadership.* How the behavior of all leaders within the project inspires, supports, and promotes project excellence
- *People.* How project team members are involved and how their potential is seen and utilized
- *Resources.* How existing resources are used effectively and efficiently
- *Processes.* How important project processes (content and management processes) are identified, checked, and changed, if necessary

The criteria for the assessment of project results include the following:

- *Customer results.* What the project achieves regarding customer expectations and satisfaction
- *People results.* What the project achieves concerning expectations and satisfaction of the employees involved
- *Results of other parties involved.* What the project achieves concerning expectations and satisfaction of other stakeholders involved
- *Key performance and project results.* What the project achieves regarding the expected project results

All criteria are further described. The criteria for project objectives of the section project management is shown in Table 7.5 as an example.

Table 7.6 shows how the processes regarding project management can be evaluated as excellent. A similar table is used for evaluating how far the project results are excellent.

Process of Assessment and Methods

The project excellence model may be applied in project reviews and evaluation and is quite commonly applied, for instance, in Germany for internal use as a self-assessment and pos-

TABLE 7.5. CRITERIA: PROJECT OBJECTIVES.

Criteria: Project Objectives (140 points)

How the project formulates, develops, checks, and realizes its objectives:
1.1. Application and demands of parties involved are identified.
1.2. Project objectives are developed, as well as how competitive interests are integrated.
1.3. Project objectives are imparted, realized, checked, and adapted.

TABLE 7.6. ASSESSMENT TABLE FOR PROJECT MANAGEMENT.

Sound Process	Systems and Preventions	Checking	Sophistication and Improvement of Business Effectiveness	Integration into the Normal Project Work and Planning	Model for Other Projects	Evaluation
Clear and extensive proof	Clear and extensive proof	Frequently and regularly checked	Clear and extensive proof	Perfectly integrated	Could be an example	100%
Clear proof	Clear proof	Frequently checked	Clear proof	Very well integrated	—	75%
Proof	Proof	Occasionally checked	Proof	Well-integrated	—	50%
Some proof	Some proof	Rarely checked	Some proof	Partly integrated	—	25%
		No proof				0%

sibility for reflection by the project team. Based on the self-assessment results, steps for further improvement are taken.

If a project team applies for the International Project Management Award, the process is as such:

1. *Application.* A project team applies for with a written statement explaining how it fulfils the criteria of the model for "Project Excellence." The statement is a self-assessment that helps the project team to understand how to achieve success and to identify and use their strengths and improvement potential.
2. *Assessment.* The assessment of the application is done by a team of at least four assessors with a high amount of project management competence. The assessor teams have different nationalities and different project backgrounds. In the first step the assessors do their assessments on their own. The lead assessor collects the single results. The criteria where the assessors have different opinions have to be discussed in a consensus meeting. The assessor has to find a consensus on the results and give recommendation to the jury regarding whether the project should enter the second stage of the assessment. The jury has the final decision on which of the candidate projects gets a site visit.
3. *Site visit.* The site visit takes one to two days, where the assessors conduct interviews with project representatives, as well as observations and documentation analysis. The assessors conclude their assessment report and again give a recommendation to the jury.
4. *Jury decision and award ceremony.* There are three categories of prizes. All the projects who got a site visit are finalists. Among the finalists, the jury decides the Prize Winners; one of these can be selected as Award Winner. Not every year an Award Winner is selected.
5. *Assessment report.* Finally, all candidate projects get their assessment report, which shows them strengths and areas of improvement.

Limits of the Project Excellence Model

There is no doubt that the project excellence model is a good project management quality management tool, as it provides a clear link between the quality of processes and the quality of results. Nevertheless, there are limitations to the model. The assessment has to stay at a high level, since it is a generic model applicable to any type of project. But for the project management section, it would be possible to go one step further and include the choice of a project management approach as the basis for the assessment. Currently, no specific PM method has to be applied by the candidate project. For instance, whether or not the stakeholder analysis is used may not matter as long as the project can prove that some kind of structured method is used to analyze their stakeholders. This kind of criticism is, however, inherent in excellence models, because they are non-normative. Nevertheless, the project excellence model has proven to be a useful project management quality management method.

Quality Management in the Project-Oriented Company

For all types of projects and programs, one can summarize that quality management has to be an integral part of the contents and management processes. The quality management of

TABLE 7.7A. OVERVIEW OF QUALITY MANAGEMENT METHODS IN THE POC.

	Objective	Standards	Methods	Roles
Project audit and review	To assess • Project the quality of deliverables and/or • The contents process	• Process and product standards	• Different combinations possible	• Representatives of the project or program • Auditor/reviewer • Audit/review owner
Management audits and reviews	To assess • the management competence of the project	• Project management and program management standard • Standards for Management audits and reviews	• Different combinations possible: • Interviews, documentation analysis, observations, self-assessments	• Representatives of the project or program • PM auditor/reviewer • PM audit/review owner

Project Excellence Model	To assess
	 • the excellence of the project management process and the content processes
	 • the excellence of the project results
		• The model is the standard.
		 • Reference to ICB, but no further standards.
			• Self assessment
			     or
			 • Assessment for award
			 • Combination with benchmarking possible
				• Project or program team
				     or
				 • Project or program team
				 • Assessors
				 • Jury

Project and program benchmarking	To assess and compare • the project performance and or • the content processes and or • the project management process and it results	• Maturity models • Best practices • Performance criteria • Process models • Project management standards	• Different step models possible	• Benchmarking institution • Benchmarking partners
Project and program consulting	To support • The project or program in a complex situation	• Project management and program management standard	• Different combinations possible: Interviews, documentation analysis, observations, self-assessments • Facilitation of workshops	• Representatives of the project or program • PM consultant

TABLE 7.7B. OVERVIEW OF QUALITY MANAGEMENT METHODS IN THE POC.

	Objective	Standards	Methods	Roles
Project manager or program manager coaching	• To support • the project manager or program manager in a complex situation	• Project management and program management standard	• Different combinations possible: Interviews, documentation analysis, observations, self-assessments	• Project manager or program manager • PM coach
Project or program evaluation	• To assess the • project performance and the management process • after the project or program is completed	• Project performance criteria • Project management and program management standards	• Different combinations possible: Interviews, documentation analysis, observations, self-assessments	• Evaluator • Project team
ISO certification	• To assess and certify • The quality of the processes of a company or company unit • By a third party	• ISO standards	• According to ISO, e.g., ISO audits	• Certification body • Auditors • Candidate
PM Certification	• To assess and certify • the PM competence of project management personnel • By a third party	• IPMA Certification: ICB, NCB • PMI Certification: PMBOK	• According to PM certification system	• PM certification body • Assessor(s) • Candidate
Accreditation	• To assess and accredit • a product, methods or processes • by a third party	• Accreditation standards	• According to accreditation procedure	• Accreditation body • Surveyor(s) • Candidate

the project or program further depends on the quality management of the project-oriented company. Project-oriented companies often have a quality management system based on a combination of ISO certification and excellence models and use different quality management methods to continuously improve their performance. The challenge is to integrate process management, quality management, and project management as specified in the new ISO 9000:2000 standards.

Table 7.7 summarizes the quality management methods applied in the project-oriented company. ISO Certification and accreditation is mainly used at the level of the project-oriented company or the line unit. They are both relevant for the project and program; for instance, if the company is ISO-certified, its projects and programs have to run according to ISO processes.

Within the project-oriented company, the PM office has an important role regarding quality management in projects and programs. Many PM offices provide quality assurance services like management audits and reviews, management consulting of project and programs, coaching of project managers and program managers, and project and program evaluation. The PM office is responsible for the management processes of project and programs and the standards and guidelines to ensure the quality. (See the chapter by Powell and Young.) Often the PM office supports PM certification of PM personnel and is responsible for other human resource functions, as competent PM personnel is also an issue of quality management (see the chapter on HR by Huemann, Turner, and Keegan). The reason for making the PM office responsible for project and program management quality assurance in the project-oriented company is that improvements can be implemented and communicated faster than when the task is left to the single project or program (Brucero, 2003).

One major thing I learned from the project-oriented company Transsystem, situated in the middle of nowhere, as they stated themselves: You can be professional and deliver high-quality solutions no matter where your company is located (Stroka, 2002).

References and Further Reading

Anbari, F. T. 2003. An integrated view of the six sigma management method and project management. In *Project Oriented Business and Society* 17th IPMA World Congress on Project Management. June 4–2, Moscow.

Bounds, G. M., G. H. Dobbins, and O. S. Fowler. 1995. *Management: A total quality perspective*. Cincinnati: South-Western.

Bucero, A. 2003. Implementing the project office: Case study. In *Creating the project office: A manager's guide to leading organizational change.* L. Randall, R. J. Graham, and P. C. Dinsmore. New York: Wiley.

Champ, R. C. 1989. *Benchmarking: The search for the industry best practices that lead to superior performance.* Milwaukee, WI: ASQC Quality Press.

Corbin, D., R. Cox, R. Hamerly, and K. Knight. 2001. Project management of project reviews. *PM Network* (March)

Crosby, P.B. 1979. *Quality is free.* New York: McGraw-Hill.

Davenport, T. 1993. *Process innovation.* Boston: Harvard Business Press.

Deming, W. E. 1992. *Out of the crisis.* Cambridge, MA: MIT

Feigenbaum, A. V. 1991. *Total quality control.* 3rd ed. New York: McGraw-Hill.

Frieβ, P. M.,(1999. *Projekt Management für den tiefgreifenden organisatorischen Wandel mittelgroβer Einheiten.* Bremer Schriften zu Betriebstechnik und Arbeitswissenschaften, Band 25, Verlag Mainz, Wissenschaftsverlag Aachen.

Gareis, R., and M. Huemann. 2003. Project management competences in the project-oriented company. In *People in Project Management.*, ed. J. R. Turner. Aldershot, UK: Gower.

Gareis, R. ed. 2002. *pm basline: Knowledge elements for project and programme management and for the management of project-oriented organisations.* Projekt Management Austria, www.p-m-a.at/ publikationen.htm.

Huemann, M. 2002. *Individuelle Projektmanagement Kompetenzen in Projektorientierten Unternehmen.* Europäische Hochschulschriften, Peter Lang Verlag, Frankfurt-am-Main.

Huemann, M., and R. Hayes. 2003. Management audits of projects and programs a learning instrument. In *Project oriented business and society.* 17th IPMA World Congress on Project Management. June 4–6, Moscow.

Imai, M. 1986. *Kaizen: The key to Japan's competitive success.* New York: McGraw Hill.

Ishikawa, K., and D. Lu. 1985. *What is quality control? The Japanese way.* Englewood Cliffs, NJ: Prentice Hall.

ISO. 2000. *ISO 9,000: Quality management systems—Fundamentals and vocabulary.* Geneva: International Standards Organization.

———. 2000. *ISO 9,001: Quality management systems—Requirements.* Geneva: International Standards Organization.

———. 2000. *ISO 9,004: Quality management systems—Guidelines for performance improvement.* Geneva: International Standards Organization.

———. 2002. *ISO 19,011: Guidelines for quality and/or environmental management systems auditing.* Geneva: International Standards Organization.

———. 2003. *ISO 10,006: Quality management—Guidelines to quality management in projects.* Geneva: International Standards Organization.

Juran, J. M. 1950. *Quality control handbook,* New York: McGraw-Hill.

———. J. M. 1986. *Out of the crisis. Cambridge,* MA: MIT.

Logothetis, N., and H. P. Wynn. 1989. *Quality through design: Experimental design, off-line quality control and Taguchi's contributions.* Oxford, UK: Oxford University Press.

Morris, P. W. G. 1999. Key issues in project management. In *The Project Management Institute project management handbook,* ed. J. K. Pinto. San Francisco: Jossey- Bass.

OGC. 2002. *Managing successful projects with PRINCE2.* 3rd ed. London: The Stationery Office.

Pharro, R. 2002. Processes and procedures. In *The Gower handbook of project management,* ed. J. R. Turner and S. J. Simister. Aldershot, UK: Gower.

Pinto, J. K. 1999. Managing information systems projects: Regaining control of a runaway train. In *Managing business by projects. Proceedings of the NORDNET Symposium,* ed. K. A. Arrto, K. Kähkönen, and K. Koskinnen. Helsinki: Helsinki University of Technology.

PMI 2000. *A guide to the project management body of knowledge.* Newtown Square, PA: Project Management Institute.

Roy, R. K. 1990. *A primer on the Taguchi method.* New York: Van Nostrand Reinhold.

Scott, B., ed. 2001. *Partnering in Europe: Incentive based alliancing for projects* London: Thomas Telford.

Seaver, M., ed. 2003. *Gower handbook of quality management.* Aldershot, UK: Gower.

Senge, P. 1994., *The fifth discipline fieldbook: Strategies and tools for building a learning organization.* New York: Doubleday.

Sroka, S. 2002. Reorganisation of transsystem to a POC for the re-inforcement of the customer focused market. In *Making the vision work. 16th IPMA World Congress on Project Management.* Berlin, June 4–6.

Stummer, M., and M. Huemann. 2002. Development of competences as a project-oriented company: A case study in the chemical industry. In *Making the vision work.* 16th IPMA World Congress on Project Management. June 4–6, Berlin.

Tennant, G. 2002. *Design for Six Sigma, Launching New Products and Services without Failure*, Aldershot, UK: Gower.

Turner, J. R. 1999. *The handbook of project based management. 2nd ed.* London: McGraw-Hill.

———. 2000. Managing quality. In *Gower handbook of project management*, ed. J. R. Turner and S. J. Simister. Aldershot, UK: Gower.

———. 2003. Farsighted project contract management. In *Contracting for project management*, ed. J. R. Turner. Aldershot, UK: Gower.

Van Wieren, H. D. 2002. Alliance, an excellent solution to meet project execution challenges. In *Making the vision work*. 16th IPMA World Congress on Project Management June 4–6, Berlin.

Wateridge J. 2000. Project health checks. In *The Gower Handbook of Project Management*, ed. J. R. Turner and S. J. Simister. Aldershot, UK: Gower.

Webb, A. 2000. *Project management for successful product innovation*. 2nd ed. Aldershot, UK: Gower.

CHAPTER EIGHT

THE PROJECT MANAGEMENT SUPPORT OFFICE

Martin Powell, James Young

As the practice of project management has grown, so has the demand for a systematic method of implementation. In the late 1970s, the establishment of a project support office was seen as the vehicle to achieve this and was traditionally responsible for status reporting. It was staffed by project management professionals serving the organization's needs through the provision of support. From this project office or project support office stemmed the project management support office (PMSO), combining the duties of supporting projects and reporting project status with a Project Management Centre of Excellence.

The PMSO is a central organizational unit that is responsible for ensuring the portfolio of projects performs optimally. It does this through the provision of support to portfolio, program, and project managers; the creation of a project management function (this is often "virtual," i.e., it has no functional mandate but provides a "home" for project management); the development of management competencies appropriate for the management of the portfolio; and the provision of effective tools.

For project team members, the PMSO should be an invaluable source of support. The PMSO can supply team members with mentors, facilitators, and just-in-time support. It offers guidance on project management methodology, standards, and processes and can provide team members with the tools they need to do their jobs effectively. It can maintain a library of previous project management solutions that might be reusable on new projects, be the owner of project knowledge, be responsible for communication, be a source of topic expertise, provide audit and "health check" support, and be a "home" for the project management community.

The roles that a PMSO performs are broad ranging—they need to reflect the needs of the organization, its structure, the type of projects it undertakes, and, in particular, its culture. A PMSO often comes into its own in a "virtual" organization—one where the company's operations and projects are conducted across a number of locations—where communication is stretched, standardization is difficult to achieve, and support is difficult to deliver. What is also true is that implementing a PMSO in a virtual organization presents a far greater challenge, both organizationally and culturally.

The Need for a PMSO

Every day, organizations commit enormous quantities of resources to projects. Not all organizations can be certain that they have sufficient data at hand to ensure that they are investing in the right projects, that the business case for the projects they have chosen to invest in remain valid, or that their current portfolio of projects will deliver the required product within the time and cost resources specified. The role of the PMSO is to support management by validating that the information that is received is accurate and that those they put in charge of delivering their portfolio are properly equipped to do so.

Not everyone embraces the idea that a PMSO adds value to the project management capabilities of an organization. Many may view it as yet another layer of bureaucracy that reduces the agility of an organization's project management function. Others see it as an attempt to stifle innovation, expose them to audit, and diminish their ability to manage their projects the way that they see fit and put them under the watchful eye of a centralized control group.

Organizations that wish to establish a PMSO should realize that they might encounter substantial skepticism and resistance to their efforts. It is important at the outset that well-defined steps are taken in educating individuals about the need for a PMSO and that the way it will function and its deliverables are clearly communicated. To be successful, a PMSO must be supported by the project management community; the community must feel as though the PMSO is there to provide support and guidance it when required.

A good starting point is to identify what questions might be raised about a PMSO. A PMSO must be able to define its value to projects, define what it is going to produce and why it is producing it, and be able to understand the impact of its work.

The aim of a PMSO must ultimately be to improve the performance of projects in an organization. How this is measured depends on the context in which the organization operates—but it always has to deliver greater "benefit."

Purpose

The objectives of a PMSO are a key factor in determining its structure, responsibilities, and organizational "location." Usually the principle objective of establishing a PMSO is to improve the overall performance of projects within an organization. But what does this mean? In some companies this is viewed as ensuring projects deliver effectively within the classic time, cost, and quality criteria; in others it can mean improving the performance of the portfolio and its impact on the business.

The objectives often reflect the sector or "orientation" of the organization; contracting companies—for example, construction or IT service organizations—often seek to use the

PMSO to improve the competence and quality of their project managers and thereby the service they offer to their clients. In contrast, product development companies—for example, software vendors or pharmaceuticals companies—often mandate the PMSO with improving the benefit that the portfolio of projects delivers back to the business in terms of return on investment.

The principal differences between these two models are the scope of the PMSO's responsibilities, the mandate that they are handed, and its location within the organization.

Structure

The structure of the PMSO depends upon the scope of its responsibilities, its location within the organization (does it provide support at project, program, and/or portfolio), and the size and structure of the organization it serves. The PMSO can exists at three levels (Crawford, 2002):

- *The strategic PMSO.* Responsible for setting corporate policies and standards for projects, for coordinating their implementation down through the organization, and for supporting portfolio managers in aligning the project portfolio with business strategy. The strategic PMSO is also responsible for ensuring effective communication between business unit PMSOs to facilitate the sharing of best practice and the transfer of knowledge.
- *The business unit PMSO.* A dedicated project office responsible for implementing the standards and policies established by the strategic PMSO within a specific business function or unit. Often corporate standards need interpreting to the specific needs of a business unit (BU) or function. While the principles remain the same, the specific processes, templates, and control standards may vary depending on the types of projects under way. The BU PMSO might also be responsible for managing project resources and for ensuring the effective transfer of knowledge between projects within the BU.
- *The project office.* Large, complex projects might warrant the existence of a dedicated project office organization. This office is responsible for ensuring the standards and policies established by the strategic PMSO and BU PMSO are applied. It is also often responsible for the implementation of information management standards, as well as the hands-on management of the project control function.

In fact, these offices do not need to be separate entities—in some organizations one PMSO might undertake the roles of strategic PMSO, BU PMSO, and project office.

Mandate

The roles and responsibilities of the PMSO must be clearly established from the very outset. This includes direction on which of the PMSO's services, standards, and policies are to be "imposed" on projects—mandatory—and which should be made available to projects should they be required.

FIGURE 8.1. LOCATIONS OF THE PMSO.

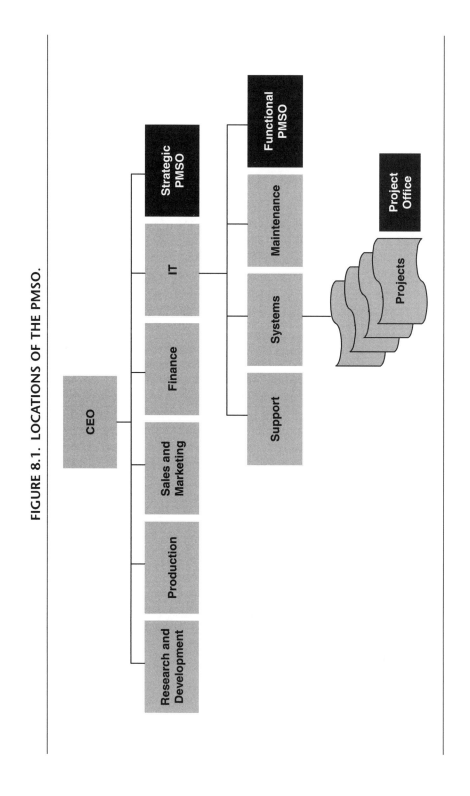

Services that are mandatory must be determined by the sponsor of the PMSO but must reflect the structure and culture of the organization—a devolved organization operating in autonomous business units often rejects "interference" from a central organization, while centralized organizations tend to be far more accepting of company-wide standards.

There are no hard-and-fast rules on which standards, policies, and services should be mandatory; however, those that are mandated typically tend to focus on the use of the PMSO to provide reviews of projects at key milestones and standard structuring, recording, and reporting of project information—in order that it can be easily aggregated to program and portfolio level.

Another service that can be mandated is the use of the PMSO to audit projects. Where this is done, care should be taken to ensure that the role of the PMSO, to provide objective support to project managers, is not compromised by its role in reviewing their performance.

Sponsorship

During PMSO implementation, a critical factor is the appointment of a sponsor at an appropriately senior level. The role of the sponsor is to ensure that the deliverables, in this case the services of the PMSO, remain in line with the needs of the organization and to provide it with strategic direction.

The sponsor of the PMSO has some basic roles: to support, promote and provide strategic direction to the PMSO, as well as determining its responsibilities, structure, and mandate. The sponsor for each of the three levels of the PMSO can be different, but the basic roles remain the same.

Strategic PMSO

The sponsor must be a senior manager within the organization who has a vested interest in the PMSO making an impact on the performance of the business. Because the strategic PMSO often lies across several business units or functions, the sponsor's role may often be undertaken by a sponsoring board. The board will be made up of senior representatives from each of the functions or business units covered by the PMSO.

Business Unit/Functional PMSO

The sponsor is usually a senior manager, often the head of the business unit or function, who requires an improvement in the performance of projects and greater certainty of outcome of projects within their organization.

Project Office

The sponsor is usually the project manager who requires a project control function with added responsibility for ensuring that corporate standards and policies are being adhered to. Whoever the sponsor is, they must be willing to actively communicate the roles and responsibilities of the PMSO (at any level) and to champion its use throughout the organization.

Charging for PMSO Services

Charging projects for the use of the PMSO's services is often a delicate matter; a balance needs to be struck between encouraging the project community to engage with the PMSO in order to embed best practice across all projects and ensuring that the PMSO continues to provide tangible value-added services to projects.

Studies have indicated that there are mixed feelings about charging for any PMSO services, although generally it is accepted that charging might be appropriate for "consultancy"-type services—for example, project ramp-up support, workshop facilitation, training, and project "health checks."

One such study (Morris et al., 2000) found that charging for these types of service provided some benefit to the PMSO:

- It helps prevent the businesses and projects from using the PMSO for trivial, nonmandatory, or non-critical tasks.
- It encourages the PMSO to meet some of its budget through the sale of its services to projects. This encourages the PMSO to ensure that the services it provides are cost-effective and relevant to the business.
- It enables the PMSO to raise its profile across the organization, helping to attract high-caliber project personnel to work within it.

Responsibilities of the PMSO

The responsibilities of the PMSO are split between the different levels described earlier in this chapter. The chart in Figure 8.2 identifies the responsibilities that each of the offices might reasonably take.

The strategic PMSO role focuses on supporting the portfolio management role, defining best practice, facilitating communication across the project community, and collecting and reporting the status of all projects within the portfolio. BU PMSOs focus on the implementation of best practice in the context of their specific organization of function, while ensuring that project information is reported consistently in accordance with standards set by the strategic PMSO. The project office is concerned with applying the policies and standards at a micro level on specific projects.

The responsibilities of the PMSO fall into four categories:

- *Portfolio, Program, and Project Management Support.* Where the PMSO provides support directly to portfolio managers, program managers, and project teams
- *Organization.* Concerned with the creation and management of the program and project management community
- *Competence.* Identification, structuring, and delivery of knowledge, including training and communities of practice
- *Systems and Tools.* Designing, training, and supporting the tools and systems used to support portfolio, program, and project managers

FIGURE 8.2. RESPONSIBILITIES OF THE THREE LEVELS OF PMSO.

Responsibilities	Strategic PMSO	Business Unit PMSO	Project Office
Portfolio, Program, and Project Support			
Portfolio management support	✓	✓	
Project health checks and audits		✓	
Project governance and reviews	✓	✓	
Reporting	✓	✓	✓
Project ramp-up support and project rescue		✓	✓
Project Management within the Organization			
Resource planning		✓	✓
Communications	✓	✓	
Benchmarking	✓	✓	✓
Performance measures and metrics	✓	✓	
Competency			
Best-practice guidance	✓	✓	
Tailored methodologies	✓	✓	
Communities of practice	✓	✓	
Training and building capability	✓	✓	
Coaches and mentors	✓	✓	
Knowledge management	✓	✓	
Systems and Tools			
Project management systems design	✓	✓	
Software reviews and recommendations	✓	✓	
Design of standard templates and documents	✓	✓	✓
Design of procedures	✓	✓	✓

Portfolio, Program, and Project Support

The PMSO is responsible for providing support to all levels of the business on the status of projects:

- Strategic PMSO provides support to the portfolio management function on the current status of its cadre of programs and projects, while informing BU PMSOs of the standards that it must adhere to.
- BU PMSOs provide support to the strategic PMSO and the function or business unit in which they operate in determining the current status of programs and projects, as well as establishing and communicating the standards that programs, projects and project offices must adhere to.

- Project offices provide support to program and project managers to determine the current status of the work that they are responsible for, as well as structuring data for aggregation by the BU PMSO and/or strategic PMSO.

Portfolio Management Support

The strategic PMSO can provide support to the portfolio management function: its primary function in this area is the acquisition, structuring, and reporting of portfolio-level information from programs and projects. Key to this process is the establishment of data standards—defining which information is required (this should include an agreed definition of terms such as "accrual" and how often it is required.) The strategic PMSO should define these standards and ensure that they are communicated to the functional PMSOs and the program and project managers.

The strategic PMSO should ensure that the data in the portfolio system is regularly updated in accordance with the process agreed by portfolio managers and that the data submitted is accurate. The information stored in the portfolio system must be the same as that in the project control system, although it might not be updated as frequently. Portfolio reports should be automatically generated from the portfolio management system.

In addition to this, the strategic PMSO might also provide assistance in applying different scenarios into the portfolio system to demonstrate the impact on the performance of the portfolio of changes to project cost, time-to-market, and project revenues.

Project Health Checks and Audits

The provision of project health checks takes on two forms: audit and self-assessment. Auditing is typically seen as a policing role deemed mandatory for projects. An audit consists of a formal review of the project against a set of quality criteria. It focuses on compliance as well as providing a snapshot status review of the project for the benefit of the sponsor. Often the competency of the staff is also reviewed and reported to the sponsor.

The other form of health check is self-assessment, where the PMSO provides the project manager with the tools and training to enable him or her to assess the performance and compliance of his or her own project. Self-assessment of projects is rarely deemed to be mandatory, but it is used to encourage project managers to revisit their projects prior to major gates or milestones.

The PMSO can face a dilemma if project audits are mandated; a balance must be struck between the PMSO's role as the provider of knowledge and support—demand-driven services—and audit. Projects might be less inclined to discuss issues, problems, failures, or their support requirements if they feel that their performance is to be judged by that same organization during an audit. This is particularly true if the outcome of audits have any bearing on the way project managers are assessed for career progression.

Project Governance and Reviews

The PMSO (strategic or BU) can assist the businesses manage their projects through the governance process. The role of governance is to monitor project progress against budget and milestones; to ensure that the business case remains valid; to provide strategic direction on issues and changes; to ensure that corporate standards, processes, and procedures are being properly applied; and to facilitate the removal of organizational blockages to successful project completion.

A project's governance board is made up of project stakeholders—people with a vested interest in the project's execution and delivery. A PMSO often has representation on the governance board with a specific focus on the application of good project management and to ensure that progress information provided is accurate. The PMSO can provide advice to project managers regarding how to address shortfalls in the application of project processes and where support may be sought.

Reporting

The PMSO is often responsible for generating status reports. The first stage in this process is to gather, structure, and validate progress data. The data collected and reports generated vary by level in the organization:

- The strategic PMSO collects portfolio-level progress information, typically, completion date, forecast cost at completion, and updated business case. It then report this to investment committees and business planning.
- BU PMSOs collect status information on programs, projects, and the BU portfolio typically performance against budget and baseline, as well as updated business case. Reports are generated for the business unit managers, and the data is structured to allow it to be aggregated by the strategic PMSO into its portfolio management systems.
- Project offices are responsible for acquiring updated information on the progress of the program or project that they serve. Reports are generated for the program/project manager, the project sponsor, and any other party that requires project information. In addition, the project office structures its information in such a way as to allow it to be aggregated by the BU PMSO and the strategic PMSO.

Project Ramp-up Support and Project Rescue

The PMSO often provides the traditional project office role of project control. The PMSO can either provide a resource to fulfil the project control role or provide the project control service to the project as a clearly defined set of deliverables.

PMSOs might retain the services of some high-performing project managers that can be deployed rapidly on projects, temporarily, either to rescue projects that are failing or to ensure that projects are structured correctly from the outset.

Where the PMSO is called in to rescue a project, its first task is to assess the situation and to provide a clear strategy on what actions must be taken to rescue it, including the implications to time, cost, and so on. Where a project has encountered significant problems or appears to be badly failing, a judgment on whether to continue the project must be made with the sponsor.

The PMSO's services for supporting the set up of projects are often used on particularly complex or large projects where highly competent and experienced project personnel are required to ensure the project is correctly structured, or where the project is unable to staff itself with suitable resources in the immediate term. The PMSO is responsible for ensuring that the project is established in accordance with the best practices that it has established.

Where these types of services are provided, the PMSO often charges the project for its project managers' time on a consultancy basis.

Project Management within the Organization

The following section deals with the responsibilities of the PMSO on an organization-wide basis. It addresses the role undertaken in planning project management resources across business units and functions and how these resources are managed. This section also addresses how the PMSO communicates with the organization and how benchmarking should be undertaken to ensure that project practices are effective.

Resource Planning

The PMSO can take on a number of roles with regard to the management of the project resource pool. It can have full responsibility for the day-to-day management of project managers at one end of the spectrum, and at the other it can have no responsibility at all. It can also take on a myriad of roles between these two scenarios.

In organizations where project management is a central function, the PMSO undertakes the same roles as any other corporate function. As well as the traditional "hiring and firing" roles, it might also include the identification of skill requirements (competencies); forecasting the number and types of resource that are needed to manage the forthcoming portfolio of projects; monitoring actual resource use; providing guidance, mentoring, and training to project managers; conducting performance reviews; and monitoring career progression.

Where the PMSO is responsible for forecasting resource requirements it must liase closely with the portfolio management function to plan the number of project managers required in the future and to ensure that the right caliber and seniority of project manager is available to manage the up-and-coming portfolio of projects. Coordination between the business unit PMSOs is often required to meet peaks in demand across the organization.

The PMSO might be responsible for managing the whole project management resource pool or might retain a few high-caliber individuals in a central pool. In the latter case, these project managers might be deployed to manage high-profile, complex, or very large projects, or might be used to undertake project audits, sit on project review boards, provide project ramp-up support, or step in to rescue projects that are in trouble.

A *project management competency framework* is often used in mature PMSOs to ascertain the project management skill sets across the organization. The competency framework is also useful in identifying skill gaps across the organization and allows the PMSO to focus its training to ensure that the skill needs are met across the organization and resource can be deployed cross-functionally if required.

Where the PMSO is not responsible for the management of the project management resources pool it often maintains a database of project managers and their skill sets in order that project managers can be redeployed across the organization where required.

Communications

The PMSO should also establish and execute a strategy for communicating with the project management community. This strategy should determine what messages it intends to broadcast—usually updates on services offered, news, access to information, knowledge, training and contact details, and its target audience, including program managers, project managers, project team members, project sponsors, business managers, and the media that it intends to use to communicate.

In recent years Internet technologies (Internet, intranet, and extranet) have provided a highly effective medium for supporting communication strategies. More advanced technologies can be deployed to aggregate information on progress, status, performance, or any other measure, as well as reference and support information. This is discussed later in this chapter under *Portals*.

To support its role as "competency developer" the PMSO should play a role in advertising project management courses within the organization and generating course attendance. The PMSO also has a major responsibility to monitor the quality of the training efforts and to provide the guidance needed to deliver first-rate course offerings.

The methods of communication can vary widely depending upon the functions of the PMSO. Many organizations communicate through e-mail notifications, but other forms of media such as posters, calendars, summary sheets, and intranet sites promoting the products and services are also used.

Benchmarking

For a company to assess its performance, it must compare itself against a reference point. This process is known as *benchmarking*, and it is critical in the management of any organization. The PMSO should actively lead or support the process for benchmarking projects and project management.

The PMSO must begin with an assessment of the current condition of the organization with regard to its project management maturity, and then it must establish a baseline. While no industry standard yet exists for baselining the capabilities of an organization's project management functions, several models exist designed to measure project management maturity. If the maturity can be assessed, it provides the future ability to quantify the value of project management against the original baseline.

Benchmarking is undertaken for a range of reasons:

- To establish performance in comparison to competitors—to determine whether a company's project delivery capability is as effective as its rivals'.

- To determine whether the performance of the organization is improving—monitoring a metric over a period of time to determine whether improvements are being achieved.
- To establish performance by comparison with companies that have a similar focus—for example, measuring the competency of project management staff. This is often done where no industry standard is available for the metric.
- To monitor the effect of an initiative—for example, training—on the performance of projects.

Any aspect of an organization can be benchmarked—the return on investment can be compared with competitors or the competency of project managers can be compared with other companies in different sectors. Crucial to the benchmarking process is that the information being benchmarked is accurate and the population against which comparison is being made is known.

Performance Measures and Metrics

A company's portfolio of projects is the vehicle for developing new products and services or for delivering products and services to clients. Whichever it is, the company needs to know how well its portfolio is performing.

The PMSO should be involved in determining which measures to put in place, and once established, then it should be responsible for ensuring that they are effectively implemented, understood, and continuously updated on projects. The PMSO should be active in monitoring the performance of projects against these metrics, reviewing the metrics at regular intervals to ensure that projects remain on track, and helping to communicate what the metrics mean to project sponsors. The PMSO should be identifying where the metrics indicate that a project is not performing and, ideally, providing advice on corrective actions.

The measurement and monitoring of specific areas of performance allows managers to determine whether desired performance levels are being achieved or whether improvements are being made. Crucially, it also allows failures in performance to be identified and trends towards nonperformance to be identified and management actions to be taken. These measures are commonly known as "metrics."

Figure 8.3 illustrates that metrics can either be empirical—for example, *number of concurrent projects*—or subjective—for example, *client satisfaction*. They can also be classified as *input* or *output* metrics. Input metrics indicate that projects should perform well, for example, *highly competent project managers*, while output measures indicate that projects are performing well, for example, *return on investment*.

Metrics that have an empirical basis are generally more reliable (although every project manager will tell you that the numbers can be massaged) and easy to update, as the data can generally be derived from project control, finance, and HR systems. While metrics do not have to have an empirical basis, efforts must be made to ensure that subjective measures are described in sufficient detail to allow them to be applied consistently every time.

Metrics do not necessarily provide any valuable knowledge on their own. To have real meaning, they must be benchmarked. The metrics used vary according to different levels within the company: Strategic PMSOs help business and portfolio managers to focus on strategic metrics, BU PMSOs help functional and business unit managers to focus on or-

FIGURE 8.3. EXAMPLES OF METRICS BY CLASSIFICATION.

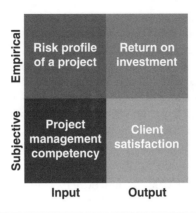

ganizational metrics, and project offices support program and project managers with project performance metrics.

The actual metrics used depends on a number of factors:

- *Whether the project organization is focused on delivering projects internally or externally.* Companies focusing on the delivery of internal projects focus more on return on investment and portfolio performance, while those delivering projects externally focus more on resource usage and margin.
- *The project management maturity of the organization.* Companies that are relatively immature find it difficult to identify information to support the calculation and update of metrics.
- *The sector that the company operates in.* The sector influences the metrics used at the program and project level to measure performance. This generally reflects industry cultures and data standards; for example, system availability is commonly used for measuring performance in the ICT sector, while earned value calculations are more common on military and construction projects.

Whichever metrics are used, perhaps the most important factor in their effectiveness is whether they are aligned to the strategic and tactical needs of the company—that is, do they tell management what they need to know? There is no correct set of metrics that should be used, but those that are implemented must be current, accurate, and repeatable.

Competency

The PMSO needs to be an ambassador for project management within the organization. It must be seen as the "voice" of how projects should be managed. To do this successfully, it needs to be the following:

- Holder of best practice guidance
- Owner of a set of methodologies
- Key driver in knowledge management initiatives
- Catalyst for communities of practice
- Locator of subject matter experts
- Key provider of training
- Possessor of up-to-date literature and institutional knowledge
- Assessor of the maturity of the organization and measure improvement
- Capable to carry out health checks and support to projects either through facilitation or general project management training
- Provider of tools to facilitate practice and also to benefit portfolio reporting

Figure 8.4 shows the PMSO as the "owner" of project or project management knowledge. This knowledge is distributed through a series of sources—guidelines, processes, procedures, and so on. Key to maintaining knowledge is for projects to implement and feedback new knowledge in the form of lessons learned or process improvements through the use of subject matter experts and communities of practice.

Best-Practice Guidance

The PMSO is the owner of best practice. It needs to hold and communicate a clear message of what project management is and how it should be carried out within the organization to optimize effectiveness. A significant factor in the success of the PMSO is its ability to effectively articulate, disseminate, and maintain a set of practices and processes that reflect how the organization expects projects to be managed. It should also identify weaknesses or areas where improvements are required and set forth a vision of where it should be developing in these areas. To achieve this success, these standards need to be organizationally specific, contextualized, concise, and, above all, practical.

The PMSO serves as a central library for these standards and is the center for expertise on their deployment. The PMSO also incorporates lessons learned on projects in project management methodologies. Clearly, lessons learned can be very specific and the methodologies should have some method of arranging learning in degrees of generality and specificity.

In a well-developed PMSO, there is usually a separate individual responsible for process development and maintenance to ensure that knowledge is used effectively. His or her role is to leverage the existing networks within the organization, or create new networks of subject matter experts who can maintain and develop specific process. The individual can also facilitate the formation of communities of practice, as discussed more fully later.

Bodies of Knowledge. As the owner of best practice, the PMSO needs to decide whether the best practice is based upon industry standards, internal company standards, or proprietary standards bought off the shelf.

FIGURE 8.4. THE PROJECT MANAGEMENT KNOWLEDGE TRANSFER.

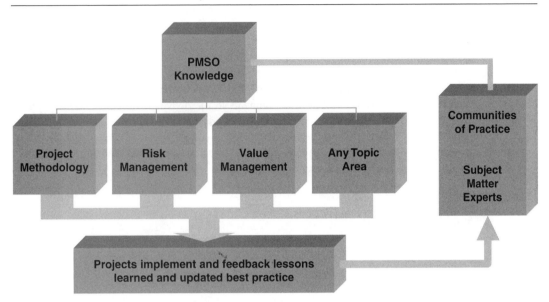

The two most significant standard guides to project management that are generally available to the public are those produced by the Project Management Institute (PMI) and the Association for Project Management (APM).

PMI's Project Management Body of Knowledge (PMBOK) provides an overall framework for thinking about project management and details nine "knowledge" areas that PMI considers are the areas "unique" to project management (PMI, 2000).

The *APM* Body of Knowledge looks at areas that research has shown can contribute to projects being successes or failures, and that therefore need to be managed. There are a total of 42 topics grouped into seven areas. (Dixon, 2000)

Generic Methodologies

In the development of a set of project management methodologies, general project management practice has to be reviewed in terms of the organization's project management needs. Topics within the established "Bodies of Knowledge—PMBOK and APM—can be evaluated in terms of an organization's life cycle or its key activities in developing its products or services. Relevant topics can then be selected to be included within the set of methodologies. More importantly, the business unit PMSOs can select topics that apply to their business area. In large organizations governance boards may mandate certain processes to be incorporated into the methodologies that the business units have selected.

A project management methodology spells out the steps to be followed for the development and implementation of a project. A sample sequence of a project management methodology starting after project approval is shown in Figure 8.5.

As the owner of these methodologies, the PMSO also maintains the templates, forms, and checklists developed to assist project managers in the delivery of their projects. In many cases the templates are an integral part of the methodologies themselves. The project management methodology may require, for example, a risk plan, and project managers are helped considerably if they can see a sample of a risk plan, including instructions for completing one. There may be several templates such as:

- The project charter
- Communication plan
- Risk log
- Issues log
- Schedule template
- Reporting template
- Resource allocation and leveling samples and matrices

These templates should be standardized across the organization, but appropriate flexibility should be built into the templates to ensure they do not stifle the team's ability to innovate. Standardized templates should facilitate the production, aggregation, and review of projects.

While the PMSO should be the owners of the methodologies, parts or all of the ownership may also exist within:

FIGURE 8.5. A SAMPLE SEQUENCE OF A PROJECT MANAGEMENT METHODOLOGY.

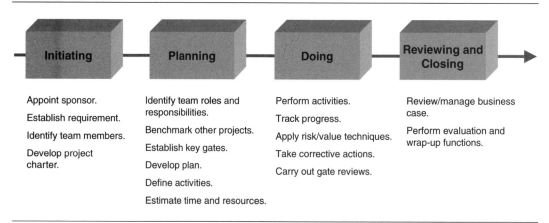

- Specialist groups or functions
- Functional PSOs
- Formal networks of practitioners
- Informal networks of practitioners
- Organizational functions (i.e., Human Resources or Finance)

This is very dependent on the organizational position of the PMSO.

Tailored Methodologies

A tailored project management methodology ensures that people and systems can speak a common language across a multiple-project enterprise setting. It must ensure that the way in which projects are carried out fits the context and the culture of the organization. It is created, and therefore "owned," by the organization, and it focuses on its specific needs—sector, culture, size, structure, and so on.

In the development of project management methodologies, a structured approach involving a series of interviews with key stakeholders and practitioners, along with a series of reviews, is necessary in order to obtain organization-wide buy-in to the proposed approach.

In large organizations this needs to be carefully planned to ensure all parts of the business are represented. To be seen as the keeper of best practice, people need to accept this proposed approach and have to make the transition from what they do now to what they need to be doing.

Figure 8.6 shows the components that make up a methodology:

- *Practices*. These are a definition of what you are trying to do.
- *Methods*. These describe how to carry out the practices.
- *Processes*. These define the sequence to how something is done.
- *Procedures*. These offer how-to detail about particular processes, methods, and practices. The methodology summarizes this detail.
- *Rules*. These are the constraints to the methodology.

These components are disseminated and presented to the organization through tools, templates, case studies, principles, lessons, and standards or standard operating procedures. The success of how this is presented to the organization is through the understanding of the needs of the people who are to apply these methodologies. This is fundamental.

Methodologies vary widely from industry to industry and from company to company. For example, Japanese product development methodologies are sharply different from U.S. software development methodologies (Rad and Levin, 2002). Global companies like IBM that do a wide range of projects, like product development, network services, manufacturing, and information technology systems integration, face the challenge of creating a common methodology that meets the needs of all project users. Once a methodology is in place, it has to be documented in the form of procedures. Procedures are the detailed how-to instructions describing the steps in the methodology, and these might need to be tailored to the needs of individual needs of different business units and functions.

FIGURE 8.6. A METHODOLOGY TRANSLATED INTO AN ORGANIZATIONAL CONTEXT.

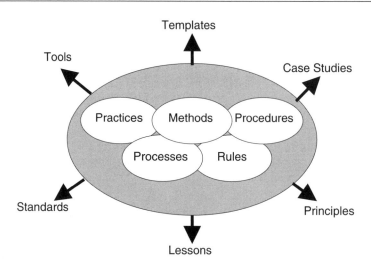

Guidelines. A set of project management guidelines offers a compromise to the methodologies. They describe at a higher level how to do project management and offer people a conceptual and more holistic view of what best practice should look like and how each project management topic should be undertaken. Where the target project management community is working across several business units, a set of guidelines would potentially achieve greater buy-in than a more detailed methodology.

Many organizations incorporate within both the guidelines or methodologies detailed information such as decision gates, toll gates, milestones, standard reporting times, portfolio management reporting, go/no-go decision criteria, and so on. This may be applied in a stepwise fashion once buy-in has been obtained to the more holistic and high-level approach.

Communities of Practice

Communities of practice are important resources, along with trainers, coaches, facilitators, mentors, line managers, and others, in ensuring that project management practice is clearly and appropriately articulated, disseminated, practiced, and kept up-to-date and relevant. *Communities of practice* are networks of interested parties who can contribute to the development and maintenance of project management best practice (Wenger, 1998). Communities can form, and operate, powerfully in organizations when focused around practice areas. In this way, groups develop a shared understanding of meaning and get improved buy-in. The PMSO may need to incentivize membership of the communities, particularly in the first one to two years, before a more steady-state community is established and membership is reliant on enthusiasm and belief.

In particular, they can be valuable in a virtual organization in using resources from the project management community (rather than relying just on a large, central PMSO) to develop guidelines and disseminate knowledge as well as facilitating the sharing of best practice and networking.

Figure 8.7 shows the communities representing a project management topic or methodology and having membership across all of the functions. Where project management is a function in the organization, this model is still applied.

Communities of practice often include people recognized as leaders or experts in one or more aspects of the practice area—for example, risk management or scheduling. These people are commonly referred to as "subject matter experts." They have an important role in articulating these project management practice areas—and in ensuring that practice is kept up-to-date and relevant. They can also

FIGURE 8.7. COMMUNITIES OF PRACTICE.

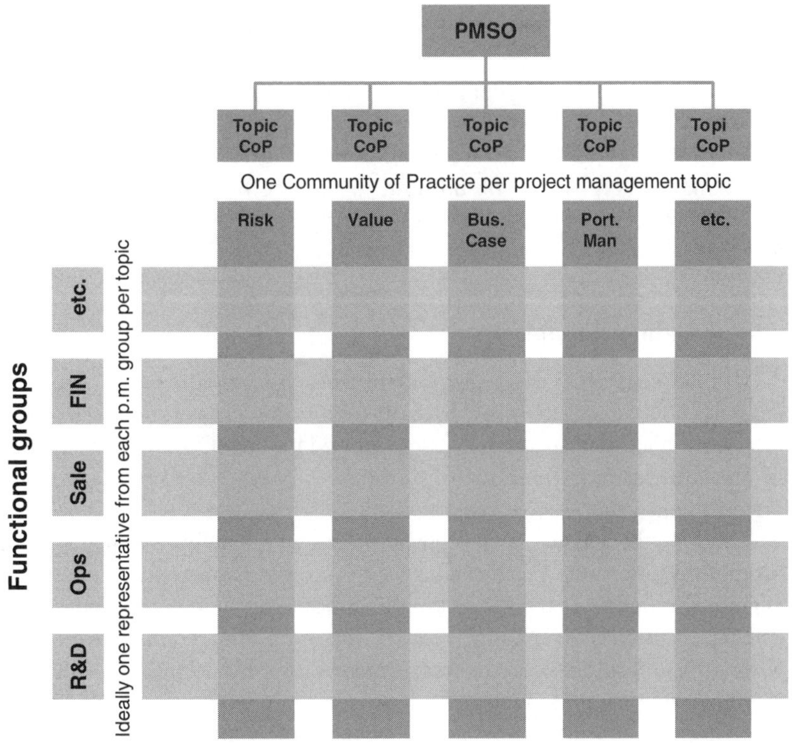

- provide help in answering questions;
- champion the topic; and
- assist in training, facilitation, coaching, or mentoring.

Communities of practice should develop as self-nominating groups as well as formally nominated groups, but as mentioned earlier, they may require incentivization at inception. Senior managers should become involved in the communities. These communities ensure that best practice remains current and promote and develop project management capability within the organization on behalf of the PMSO. The crucial task of the PMSO is to help these communities grow and to work with them so that their work is harnessed around the development of best practice for the benefit of the organization as a whole.

The level of commitment required as an individual within a community for follow-up activities, such as reviewing training material, meeting with colleagues to present practices, and acting as workshop facilitators or coaches to people taking training in the topic areas, varies depending on the following:

- Nature of the topic
- Number of people wanting assistance
- Level of effort that individuals are prepared to put in

The communities should report directly into the PMSO. This ensures that any significant changes to the current best practice can be coordinated across the entire project management community as well as across other communities of practice that may be affected by the change. There should be no limit to the size or membership of the community itself, and cross-functional or cross-business-unit participation can only strengthen the purpose of its very existence.

Training and Building Capability

The PMSO is the central focus for ensuring appropriate project management training. It identifies competencies needed by high-performing project managers, for effective executive awareness and team member participation. The PMSO participates, and typically leads, in helping tailor standardized courses around the methodologies that apply specifically to the organization.

Initially the PMSO must identify the set of capabilities that the organization needs in order to improve the performance of its project management. These capabilities can be expressed in terms of the "individual competencies"—the knowledge, skills, and behaviors needed by the project managers—and "organizational competencies"—the structure, size, geographical location(s), facilities, systems, lines of reporting, and culture in which the projects are undertaken.

The first step in improving individual competencies is to identify the knowledge and skills and behaviors that an organization's project managers should possess. This is usually done by crafting an appropriate body of knowledge framework and set of methodologies.

Once this has been done, the PMSO can then undertake a process of assessing where it actually is—that is, establishing what are the competencies and capabilities of the existing project managers. This is often done through a series of structured interviews or by a questionnaire designed to assess the project management knowledge and skill levels in the organization and also how well these are being applied.

This process allows the PMSO to identify the "gap" between where they are and where they want to be, and to set out a roadmap of how to address the shortfalls. Training courses can then be developed to address specific areas of weakness and for different maturities of project management knowledge. This form of training needs assessment can also be applied to individuals so they can determine their own training needs.

Although the training department within HR is usually the coordinator of corporate training, the PMSO may provide subject matter expertise in project management. The PMSO identifies the appropriate training that is required and may participate in the selection of the trainers. The PMSO also helps identify the required levels of knowledge and competency and the required segments of training that are necessary in order to achieve maximum performance. Thus, the PMSO is the focal point for establishing the means to measure project manager competency. Competency building can be provided in a variety of ways, and the range of offerings should be determined from the questionnaire.

A learning styles assessment can also help in ascertaining how people learn. These offerings include the following:

- *Face-to-face (F2F) classroom training.* This is typically offered in a lecture or workshop style and is designed to encompass as much of the methodology topic areas as possible. This is an ideal environment to "expose" the project management community to a wide range of topics.
- *Blended learning.* This is a combination of e-learning being supported and led by a trainer. This can be applied across different geographical locations through. The trainer controls the progress of the material.
- *e-learning.* This can be designed as pre-F2F learning or to reinforce the classroom training. As a stand-alone training package, e-learning tends to be of greater benefit where knowledge needs to be acquired rather than a skill developed. As pre-reading, it helps make the F2F training more efficient, and as a refresher to the classroom training, it helps greatly in the retention of knowledge.
- *Decision support tools.* These provide an interactive version of online learning to be viewed and used just-in-time, thus allowing practitioners to reference the training on an as-needed basis. They are designed to provide practitioners with possible solutions to real problems.
- *Facilitation (and JIT).* Perhaps the most effective form of competency development because it links the training to a real environment, this combination of applying best practice to a practical situation provides the maximum benefit and training effectiveness; people learn on the job.
- *Coaching or mentoring.* These roles promote best practice and target all practitioners on an individual basis. Strictly, coaching is more one-to-one training; mentoring is more general

advice, often of a career development nature. While they may not be structured training, they provide a means of leveraging the wealth of knowledge that exists within the organization.

The PMSO plays an important role in offering project management training to employees in the organization. The PMSO can carry out this responsibility in several ways. For example, it can work closely with the training department to develop courses that would be offered through the training department. PMSO staff could offer courses themselves, or they could identify and select outside vendors who would develop and deliver the course material. The PMSO, in every case, should play a key role in the design of the training or suite of training courses, and the selection of attendees—in order to maximize the effectiveness of the training and exploit any policy around mandatory training and optional training.

A crucial element in developing training material is how well it fits the organizational context; does it reflect the environment, culture, processes, and industry that the project managers and project team members operate in? Other than at a pretty basic level, the same training course would struggle to engage, and therefore impact, project managers working on an oil facility and those on a new financial product.

As organizations devote more resources and energy to conducting their work on a project basis, the need for project management training grows. Project staff need this training to strengthen their ability to organize and implement their work. Workers from the functional areas (such as accounting, design, marketing, purchasing, and engineering) might also need some measure of project management training to make sure their efforts mesh with the organization's project focus.

The types of training that should be offered can vary:

Project Management Basics. A course on the project management basics is geared toward project management novices or to people who just need an appreciation of projects so they can work with them more effectively. It describes the methodology and its fit with the business, people's roles and responsibilities, and some of the useful tools and techniques typically available to assist in enhancing the management function. It introduces core topics such as time, cost, and people management; the project life cycle; project management players; project politics; control; and evaluation. It would usually run from two to five days. Its principal objective is to develop an understanding of what project management is, what it does, and how it fits into the organization.

Advanced Project Management. Advanced project management may be a single course or a series of courses geared toward developing specific project management competencies, such as scheduling, cost management, or resource allocation skills. Unlike project management basics, students are expected to participate in exercises, case studies, and role-playing. The number of training days associated with an advanced project management curriculum typically ranges from 5 to 15 days.

Preparation for the Certification Examination. Many organizations accept the Project Management Institute's Project Management Body of Knowledge (PMBOK) as a useful initial basis for accreditation on project management standards. However, the APM and other country-based associations provide accreditation programs, and some companies also work with consulting companies and universities to acquire additional certification programs. There are also many public methodologies that have training offerings. The length of training depends on the standard of certificate and the overlap with other forms of training being offered.

Specialist Topics. Beyond studying the project management topics in areas like cost control and scheduling, project workers may benefit from investigating more specialized topics such as risk management, value management, and contracting and procurement (or any of the methodology topics developed by a PMSO). Not every project worker may need take these courses, which are aimed primarily toward those who wish to assume higher levels of project management responsibility. Each of these courses can be handled as a one- or two-day offering or offered as a combination of methodology topics tailored to demand, or developed in specialist cases into longer programs. These specialist topics may be undertaken as facilitated workshops; for example, value engineering may require three to five days of team effort at a key stage gate, or risk management may require three to five days over a period of weeks or months.

General Business Management. To enable project managers and other project staff to develop general business management skills, they should be encouraged to take courses that cover topics such as finance, marketing, information systems, and organizational behavior. The material can be bundled into one or two business basics courses, or it can be offered through separate specialized classes.

The need to sustain a good project management operation with training is obvious. The question that arises is this: What role can the PMSO play in developing and delivering the training material and in coordinating the overall training effort? The real issue is whether the PMSO should provide the lead role, or whether it should support the training activities through the training department. The response in most organizations seems to be that the PMSO should do some of both. On specialized topics, the PMSO often plays the lead role. On more generalist training, it plays a support role.

The PMSO plays a lead role in course development, delivery, and coordination. it must assume leadership in identifying the curriculum, because through competency assessment, as owners of best-practice and gap analyses, the PMSO staff are in the best position to identify the organization's project management training needs.

Organizations with Centralized Training Departments. It makes sense to have these departments play the lead role in developing and delivering project management courses, just as they do with other courses. Because of their centralized location, they know what the overall training program is for the entire organization and can fit the project management curriculum into the organization's broader training portfolio. This may lead to a more cohesive training effort and yield economies of scale in the delivery of training.

In this case, the PMSO gives the training staff the technical information it needs to assemble a good program. For example, if the centralized training department decides to develop its own training material, the PMSO supplies the experts who create course content.

In a sense, they are "contracted" by the training department to develop material. The training department plays the role of client, the PMSO the role of developer.

The PMSO could coordinate the supply of the instructors, since the organization's project management expertise resides there. The PMSO may assign one of its staff to work full-time with the training department throughout that period or at least until the trainers have been trained.

Organizations That Lack Centralized Training Departments. It may make sense here to have the PMSO assume a strong leadership role in developing and delivering project management courses. The PMSO must work closely with the training department in this effort. However, the roles have been reversed: Now the PMSO plays the lead role while the training office plays a support role. Support from the training office might include the following:

- Guidance on good curriculum development practice
- Historical training needs and learning styles data
- Review of course material to determine its "teachability"
- Development of standard formats for producing course material
- Assistance in the production of course handouts
- Provision of teaching technology, such as overhead projectors, LCD projectors, flip charts
- Assistance in the administration of training—organizing venues, delegates, materials, and so on.

Coaches and Mentors

Individuals assigned to look after the training needs of others (i.e., coaches and mentors) can work with the PMSO to serve a just-in-time training function. For example, when design engineers need help in learning how to capture their activities in a work breakdown structure, they can contact a coach who provides them with on-the-spot instruction on building one.

The PMSO should develop a network of coaches through the communities of practice or through interest from individuals who have undergone training. The PMSO should also ensure that mentors are sufficiently briefed about the project management offerings and needs of their staff.

The PMSO plays the lead role in supplying just-in-time training. The training department is generally not equipped to handle this kind of spontaneous response to the temporary training needs of the organization's employees. In some cases the PMSO might provide this type of support, training, facilitation, or ramp-up support in a consultancy capacity, charging for its time.

Knowledge Management

Building a world-class set of methodologies involves taking advantage of the lessons learned by project managers and the project team. An archive of lessons learned is one of the PMSO's key contributions to standardizing methodology across the organization. This library of information and data is assembled from past projects, for example:

- What worked
- What didn't work
- How it can be reapplied more effectively to benefit other projects

The PMSO may also serve as the quality audit and continuous improvement function for the project management community, since it understands what should be done in terms of methodology and training and can audit against whether or not it is being done—and if it is, whether or not it is showing value and productivity. (See the chapter by Huemann.)

The embedded knowledge and understanding of the coach, mentor or trainer, and facilitator also impacts the effectiveness of the methodologies and training. The ability for both the material and trainer to deliver both knowledge and practical solutions for implementing their practices back in the workplace has a profound impact on the value and effectiveness of training.

Furthermore, while a candidate can develop new skills during training, he or she can be lost if sufficient support is not provided back in the workplace. Learning and knowledge retention are separate activities; often just six months later, competencies are lost, despite being learned during training (Baldwin and Ford, 1988). The ability for candidates to refer back to training and knowledge once in the workplace again is key to retaining knowledge, and to the ongoing contribution of the training intervention toward improving the performance of the organization.

Systems and Tools

One of the key responsibilities of the PMSO is the definition of the tool set required to support portfolio, program and project managers, as well as to support its own activities. The term "tools" not only refers to large integrated systems; spreadsheets, Web sites, templates, and search engines are also a fundamental part of the project manager's tool set. The tools used by today's project managers can be categorized into two groups:

- Project management applications
- Project management resources, guidance and support

Project Management Applications. By project management applications we mean the systems and tools that allow project managers to collect, structure, manipulate, and interrogate information to support them in the day-to-day management of their project. Typically, these tools provide functionality to assist the project manager in the following:

- Estimating and managing cost
- Generating and distributing reports
- Recording the status of documents

- Capturing, structuring, and controlling requirements
- Recording and monitoring risks
- Tracking actions
- Communicating across the project team

For these systems to be truly effective, they need to be deployed in a consistent manner across the organization; data captured in an ad hoc manner is very difficult to structure or analyze retrospectively, and information gathered and categorized using different conventions make comparison and benchmarking difficult to perform in the future.

For this reason, many companies choose to mandate certain elements of a standard tool for use across the organization. When these tools are developed or integrated across an organization, they allow structured information to be accessed in a consistent and meaningful manner at project, program, and portfolio level. Unfortunately, this structuring often limits the flexibility with which the project manager can use the systems.

The key to developing these tool sets is not in the way that the systems interact with one another, but how intuitive the user interface is and how well the structure that the information stored in the system suits the context. For example, an oil company would need to establish different project work breakdown structures for its exploration, refining, distribution, and service station divisions, as well as corporate functions such as IT, R&D, HR, marketing, and finance. The way that project information would be structured, as well as the different terminology used, would make the sensible structuring of information in a consistent manner across the organization almost impossible, although the basics must, and can, be harmonized.

Project Management Resources, Guidance, and Support. The second category of tools covers those that provide support to the project manager in ensuring that the correct processes are applied, that proper advice or knowledge is provided at the right time, and that communication across and between projects (including the development of a virtual project management community) is facilitated.

Typically these tools provide the project manager with the ability to access the following:

- Guidelines and up-to-date good practice
- Lessons learned on previous projects—what went well and what went badly;
- The ability to contact communities of practice and subject matter experts for advice
- Access to worked examples
- Magazines and project features
- Access to the output of studies, benchmarking data, and examples from other sectors and industries

These tools allow the PMSO, as owner of best practice and guidance for project management, to deliver its knowledge and content to the project community. By providing this information in an electronic format, the PMSO can create a virtual presence across the organization without the overhead of having a physical presence across multiple locations.

This can be extremely attractive to geographically dispersed organizations. It also allows the PMSO to create and position itself at the center of a virtual project management community where issues can be discussed, problems shared, news broadcast, and information traded.

While the PMSO might not actually "own" these tools, one of its key responsibilities is to participate in the specification of the tool set functionality, selection of software, and the implementation of tools on projects.

Definition of Systems and Tools

Any organization, whether concerned with the management of projects or not, needs tools to support the exchange of information, to facilitate communication, and to assist with the collection and dissemination of knowledge. The tool set that a company deploys needs to reflect and support its people and processes—tools do not perform a function on their own, they merely support people in applying process (and in some cases force the user to comply with process).

In determining the tool set that should be deployed, the PMSO should first establish what the tools should achieve for the organization. Then the functionality of the tool set can be determined, followed by the level of integration with other systems.

Data Standards and Templates

For meaningful information to be collected and aggregated up through an organization, clear guidance must be given on what each piece of data means and how it should be calculated. The PMSO must establish these standards and ensure that they are understood and implemented vertically and horizontally throughout the organization.

A glossary of data definitions should be produced giving details on the following:

- *Data description.* What is the data and what should it be used for
- *Data definition.* How is the data identified or calculated
- *Data format.* Describe any additional information associated with the data, e.g. currency, date of production, language, and so on.
- *Frequency of update.* How often should the data be updated to reflect current status

Clear processes should also be established for the exchange of data.

Systems and Tools Functionality

Typically a project tool set deployed does some, but not necessarily all, of the following:

Provide Access to Information and Knowledge:

- Support the delivery of learning and training to users in a just-in-time manner.
- Facilitate the capture and disseminate knowledge in structured manner.

- Provide access to subject matter experts and communities of practice.
- Provide a communication platform across the project management community.
- Provide easy access to data and information (both project and functional information).

Support the Basic Program/Project Management Functions:

- Estimating and scheduling.
- Resource management.
- Cost control.
- Risk management.
- Document management.
- Action tracking.
- Communications.
- Simplify administrative tasks, such as time-sheeting, expenses, and so on.

Support the Data Capture and Reporting Processes:

- Support the capture of project information (progress, issues, risks, and so on) in a structured manner.
- Facilitate portfolio, program and project reporting plus the production of ad-hoc reporting as required.

All this is often to be achieved in an environment where the project management community is virtual—that is, geographically dispersed—and where the systems in place are designed to meet functional requirements, not project requirements.

Whether an organization implements a corporate tool set depends largely on the standardization that is mandated. The PMSO's role in defining the tool set must reflect this. The PMSO has five principle responsibilities for tools:

- Assessing the need for a tool
- Determining the functionality (requirements) of the tools, including the need for any integration with each other or with corporate systems such as HR or Finance
- Providing guidance on nonstandard tools, for example, making recommendations on which risk tools to use where none is mandated
- Supporting the effective implementation of the tool
- Providing training

Portals

One of the challenges facing project managers is being able to access the disparate range of information systems and tools that they need to support them in their everyday work. Similarly, the PMSO faces a challenge in delivering the knowledge, policies, and standards to the project community at the point of use, in a consistent manner, and in a way that can be easily accessed and referenced.

To achieve this, companies are starting to develop project "portals." As shown in Figure 8.8, portals sit across the top of disparate systems, aggregating their content to a single point. This means that from a single point, a project manager can have access to:

- Project management applications:
  - Project control
  - Risk management
  - Document management
  - Time-sheeting
  - Etc.
- Project resources, guidance, and support:
  - Knowledge
  - Guidance and best practice
  - e-learning
  - Training material and courses
  - Communities of practice and subject matter experts
  - Etc.
- Other applications:
  - E-mail
  - News
  - Functional knowledge repositories
  - Etc

Furthermore, these systems can be tailored to the individual users, so that each project manager can decide what information he or she sees depending on the project that he or she is working on, geographical location, function, or any other category. This provides a powerful tool for project managers—a single point of access to any information that they need to perform their role. At the same time it provides a vehicle for the PMSO to deliver its content and knowledge directly to project managers at their point of work. A win-win situation.

Measuring Success of the PMSO

An organization must measure the impact that the PMSO has on the performance of projects. During the start-up of the PMSO, a measure of output based on the number of guidelines or methodologies produced and the demand for training based on this output provides the PMSO with a measure of its "impact" on the organization. The PMSO can take this organizational "appetite" as a sign of success. Over time, maybe several years, the training demand can be continually assessed against the same output. The correlation can then be made against the demand for the different "levels" of training, thus providing a measure of the capability or maturity of the organization.

The impact of the PMSO may also culminate in the release of manpower in other areas. The BU PMSOs may have been duplicating efforts and no longer need to sustain

FIGURE 8.8. PORTALS INTEGRATE DISPARATE SYSTEMS AND TOOLS.

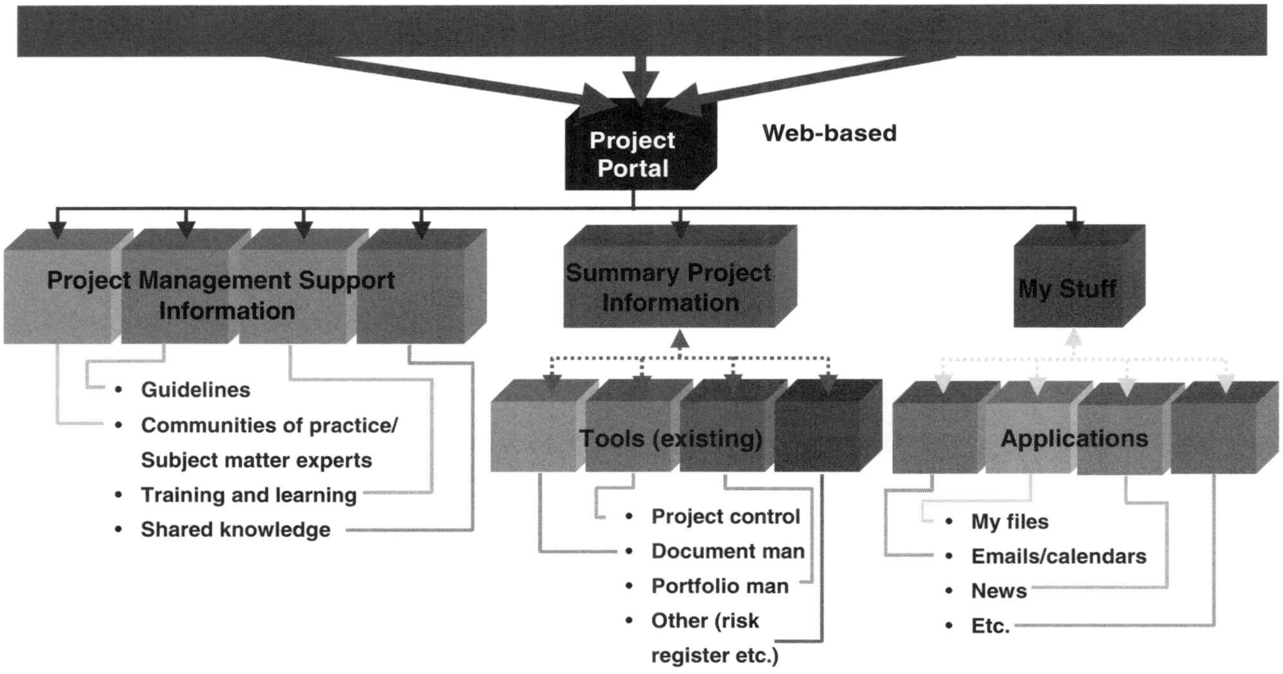

Project Portal

Web-based

Project Management Support Information
- Guidelines
- Communities of practice/ Subject matter experts
- Training and learning
- Shared knowledge

Summary Project Information

Tools (existing)
- Project control
- Document man
- Portfolio man
- Other (risk register etc.)

My Stuff

Applications
- My files
- Emails/calendars
- News
- Etc.

Source: INDECO.

their level of support to their function. It can be very difficult to attribute these changes solely to the PMSO, particularly when other organizational changes are occurring simultaneously.

Perhaps the most difficult measure of success to quantify is the direct impact that the PMSO has on the delivery of projects (i.e., reduction in project delivery timescales, costs, and so on). The PMSO should aim to demonstrate compliance to best practice across the projects and the effects it has on meeting key criteria—that is, resource utilization, meeting milestones, scope, and so on. This can be achieved be carrying out a project health check.

Finally, surveys can provide information about the perceived value added by the PMSO and the range of services that it offers. Caution should be exercised when using surveys to ensure that the data captured accurately reflects the subjective view of those responding to the survey.

Measures and Metrics

One of the ways that the performance of the PMSO can be measured is through the use of measures and metrics. These focus on specific, measurable areas that can be consistently tracked to determine whether the PMSO is meeting its targets and its impact on the organization. Commonly used metrics are as follows:

- *Indicative.* Simple measures used both during the setting up of the PMSO to monitor its productivity and as part of business-as-usual activity. These measures might include the number of guidelines produced, the number of people trained, and the number of requests for PMSO assistance.
- *Resource.* These focus on the amount of resource freed up by establishment of the PMSO by reducing the amount of rework and duplication across the organization in developing guidelines, identifying and assessing tools, and designing organizational structures.
- *Quantitative.* The impact that the PMSO has on the overall portfolio in terms of reduced time to market, reduced cost overruns, and so on. The problem with these metrics is that it is difficult to extract exactly what impact the PMSO has had in isolation of any other initiative within the organization.
- *Feedback.* "Client" satisfaction surveys to determine the perception of the impact of the PMSO. This is a highly subjective measure, and responses tend to be highly anecdotal.
- *Health.* Measuring the overall "health" of the project portfolio to determine whether projects are generally more effectively managed, whether they are hitting targets more frequently, and whether there is greater certainty in outcome.

Whichever measures and metrics are used, they must be baselined, measured, and monitored. Where possible, they should also be benchmarked to determine whether the PMSO is achieving its targets in line with industry practice.

Summary

Today's PMSO can take many forms, working at all levels of an organization, with many different service offerings and charging mechanisms. What remains true across all PMSOs is that their ultimate function is to improve the performance of projects, programs, and portfolios.

The PMSO can assume the role of mentor, auditor, supporter, facilitator, and trainer to the project management community, while its responsibilities cover the development of organizational project management competency; the hands-on support of portfolios, programs, and projects; as well as supporting the provision of the systems and tools required by project managers to effectively carry out their responsibilities.

The implementation of a PMSO can be organizationally challenging—cultural, structural, and geographical issues must all be addressed, as must the perennial barriers to change of individual inertia and organizational latency. But the benefits that a PMSO brings can be significant: Ensuring that organizations not only do projects right but also do the right projects can have a dramatic impact on overall business performance.

Whichever approach an organization takes to their implementation and ongoing management, the key to their successful establishment is to align their mandate with the strategic needs of the organization to get strong organizational support, to focus on areas of weakness and to identify itself as the natural "home" for projects and project people.

References

Baldwin, T., and J. Ford. 1988. Transfer of training: A review and directions for future research. *Personnel Psychology* 41:63–105.

Bernstein, S. 2000. Project offices in practice, *Project Management Journal* 31(4):4.

Block, T. and J. Davidson Frame. 1998. *The project office*. Menlo Park, CA: Crisp Publications Inc.

Davidson Frame J. and Block, Thomas R., 1998. *The project office* Menlo Park, CA: Crisp Publications Inc.

Franz, K-F. 2002. Crystal: Novartis' framework for IT related projects. (PMI Europe Conference, Cannes).

Dai, C. X. 2001. The role of the project management office in achieving project success. Doctoral dissertation. Washington, D.C.: George Washington University.

Dinsmore, P. C. 1999. *Winning in Business with enterprise project management*. New York: AMACOM.

Lave, J., and E. Wenger. 1991. *Situated learning*. Cambridge, UK: Cambridge University Press.

Levin, G., and P. F. Rad. 2002. *The advanced project management office*. Boca Raton, FL: CRC Press.

Ibbs, C. W., and Y. H. Kwak. 2000. Assessing project management maturity. *Project Management Journal* 31(1):32–43.

Ibbs, C. W., and J. Reginato. 2002. *Quantifying the value of project management*. Newtown Square, PA: Project Management Institute.

Kent Crawford, J., 2002. *The strategic project office*. New York: Marcel Dekker, Inc.

Kent Crawford, J., and J. Pennypacker. 2002. Put an end to project mismanagement. *Optimise* 12.

Kwak, Y.H., and C. W. Ibbs. 2000. Calculating project management's return on investment. *Project Management Journal* 31(2):38–47.

Kwak, Y. H. and C. W. Ibbs. 2002. Project management process maturity model. *Journal of Management in Engineering* 18(3):150–155.

Morris. P. W. G., and J. D. Young. 2001. Building long-term project management competencies by aligning personal competencies with organizational requirements (PMI Europe Conference, Berlin).

Morris, P. W. G., H. Khatau, J. Young, I. Keates, and J. A. Wright. 2000. Benchmarking best practice in the project support office. IPMA World Congress for Project Management, London.

Phillips J. J., T. W. Bothell, and G. Lynne Snead. 2002. *The project management scorecard: Measuring the success of project management solutions.* Oxford, UK: Butterworth-Heinemann.

Rad, P., and R. Asok. 2000. *Establishing an organizational project office.* pp. 13.1–13.9. Morgantown, WVA: AACE International Transactions, Morgantown,

Wells, Jr., W. G., and C. X. Dai. 2001. Project management offices: Organizational use is on the rise. *ESI Horizons.* (July).

Wenger, E., 1998. *Communities of practice.* Cambridge, UK: Cambridge University Press.

Young, J. D., and P. W. G. Morris. 2002. How companies are using the Internet to support their project management functions. (PMI Europe Conference, Cannes).

INDEX